The Defense Millendustry

Or
The Call for Freedom Engineers and Scientists

Or
How to Hand the (missile) Keys Over
to the Millennials (and their robots)

Jeremy Shattuck

First Stillwater River Publications Edition 2019.

Library of Congress Control Number:2019913579

ISBN-10: 1-950-339-39-4
ISBN-13: 978-1-950-33939-6

1 2 3 4 5 6 7 8 9 10

Written by Jeremy Shattuck.
Published by Stillwater River Publications, Pawtucket, RI, USA.

Publisher's Cataloging-In-Publication Data
(Prepared by The Donohue Group, Inc.)

Names: Shattuck, Jeremy, author.
    Title: The defense millendustry : or, The call for freedom engineers and scientists : or, How
        to hand the (missile) keys over to the Millennials (and their robots) / Jeremy Shattuck.
    Other Titles: Call for freedom engineers and scientists | How to hand the (missile) keys over
        to the Millennials (and their robots)
    Description: First Stillwater River Publications edition. | Pawtucket, RI, USA : Stillwater
        River Publications, 2019.
    Identifiers: ISBN 9781950339396 | ISBN 1950339394
    Subjects: LCSH: United States. Department of Defense--Officials and employees. | United
        States. Department of Defense--Personnel management. | Defense industries--United
        States--Effect of Generation Y on. | National security--United States--Effect of Genera-
        tion Y on.
    Classification: LCC UB193 .S53 2019 | DDC 355.6/10973--dc23

*The views and opinions expressed in this book are solely those of the author
and do not necessarily reflect the views and opinions of the publisher.*

Dedicated to the Freedom Engineer

Table of Contents

Definition

"Defense Industry" herein refers to the Science, Technology, Engineering, and Math (STEM) jobs in the military field, both commercial and federal, across our allied nations.

Chapter 1:
Millennials from the Trenchicles

"Each generation wants new symbols,
new people, new names"
-Jim Morrison,
Singer/Songwriter

"What do you think of our UFG...unmanned foos-scope goalie?" One night I had a dream that I installed a foosball table in my lab at our US Navy facility, an idea that was laughed about frequently among our millennial team members as a nod to what they perceived as our antithetical Silicon Valley environment. If we did have a foosball table, it would have certainly been the first at our naval base, if not illegal by our voluminous strict regulations, or at minimum, heavily frowned upon by our management. In my dream, some of my millennial team members had connected a small motor to the goalie and a control circuit that moved the

1

goalie across the table based on a proximity sensor that detected the ball and trajectory, and live streamed the match from a camera mounted on the goalie to a Periscope stream. Honestly, I had some serious reservations about the vulnerabilities of this strategy, putting non-sentient beings in charge of your defensive foos line and sharing point-of-view video of the battlefield were likely not variables that the famous war strategist Carl Von Clausewitz considered when he developed his military theories on attack and defense, but the concept was very millennial, adopting technology to automate a traditional human task and making a very near field moment for a few into a highly distributed social media event broadcast to the planet.

I myself am not a millennial, but a Gen Xer, and a Defense Industry employee since I left college twenty years ago. Over the past few years, the defense and intelligence programs I managed for the US Navy were becoming increasingly staffed and dependent upon millennials, introducing behaviors, trends, and characteristics that upset our non-millennial personnel and derail the "official guidance" processes of our programs at a daily frequency. It seemed everything that millennials did not "kill" or cancel per the popular meme, they at least disrupted, and our defense programs were not spared. Specific friction points will be covered in the following chapters, but jumping ahead to the solutions—in May 2016 I read a *Wall Street Journal* article titled "Helping Bosses Decode Millennials." The article referenced the consulting fees large Fortune 500 companies were paying for "decoding" this emerging workforce: "$20,000 an Hour!"[1] I laughed out loud imagining the US government (and taxpayer) coughing up $20K an hour of our national budget to understand what motivates their young millennial workforce and how the government could endear themselves to this young generation, as if a good paycheck signed by the world's democratic superpower was not sufficient.

In lieu of such help I was resigned to watch passionate and intelligent millennial engineers enter the Defense Industry to work on bureaucratically constrained challenges, building and integrating equipment that was obsolete by the time it left the facility, while working in multi-generational teams that had not benefited from any millennial job coaching. In fact, the only coaching conducted on the programs consisted of older team members providing history lessons to the millennials, explaining everything from cathode ray tubes, to tape drives, to Fortran, to Windows 2000. Soon thereafter I would watch that millennial passion and intensity give way to soul-sucking despair. The only silver lining was that my limited social calendar filled up with departure parties. But day by day I felt increasingly helpless; the might of the Department of Defense, the largest employer within the US government, within the USA for that matter, was unable to keep these promising engineers from making a career in a field that I had come to love, a career for me that was full of wondrous experiences.

Ultimately I did not have any other explanation as to what I was witnessing from the cubicles in the trenches of our typically mid-century government facilities (also known as the "trenchicles," and also not a selling point to millennials). So despite my laughter at the *Wall Street Journal* article, I came to realize that if companies were spending such outlandish amounts of money on understanding them, there must be something to the millennials that required significant consideration for the way the Defense Industry went about its business. Was the Defense Industry really as bureaucratic and stodgy as some of my millennial team members thought? Were we witnessing the great millennial diaspora from the Defense Industry to shiny new commercial Silicon Valley startups? Would millennials ever take ownership of the Defense Industry and sustain it?

The thing is that unlike newspapers, payphones, movie rental stores, fossil fuels, or cable TV, the Defense Industry can't be yet another item demoded by the millennial generation. The Defense Industry is a proud and historic industry; a source of great technological discoveries, iconic leaders, and there's that sort of important part where the Defense Industry maintains the national security and economy of our democratic nations! A few years ago there was a trend in the Defense Industry of re-labeling engineering positions as "Mission Engineers" to drive home the relationship of the work with the military mission, but I like to think of Defense Industry engineers as "Freedom Engineers," contributing to the protection of our democratic rights in this era of increasing threats. But with an increasing number of unfilled jobs in the Defense Industry, we need millennials to change their opinions of the Defense Industry, we need to muster millions of new millennial Freedom Engineers, and then we then need those millennials to build their own Defense Millendustry.

Chapter 2:

Who the Heck are Millennials and Why Does the Defense Industry Need Them?

"Freedom is never more than one generation away from extinction"
-Ronald Reagan,
President of the United States

Millenials were born from the mid-80s to 2000. They have no recollection of the Cold War or threat of nuclear fall-out (outside of recent North Korean threats), little concept of life before the internet or smartphones, and one of their primary cultural contributions has been the "hipster." By no means an exhaustive description, but general enough to help older folks divide the human herd. With respect to millennials in the workplace, this portion of the human herd is coming into their prime career years. In 2018 millennials became the majority of the workforce, and by 2025 they'll make up 75% of the workforce[23].

So why is there so much room in the workforce for these upstarts? Because of the departure of Gen Xers and baby boomers, who are retiring in bulk. Currently "60 percent of federal employees are over the age of 45, and nearly one-third of state and federal employees will be eligible to retire within four years."[4] This large exodus of the workforce is referred to as the "silver tsunami." Though the silver tsunami is not yet negatively impacting all segments of the economy and federal workforce, it is certainly impacting the US Defense Industry. The Defense Department wasn't quite hiring away during the blissful "peace dividend," post-Cold War era of the late 1980s to the early 2000s. In fact, from 1985 to 1995, the federal defense workforce civilian ranks decreased by over 21%[5]. Likewise, over in the private sector of the Defense Industry, cuts to large defense programs in the 1970s and 1980s as the Cold War ended kept the boomers away, and as a result, the average age for defense workers is older at 45 years old compared to the median age of 42 years old for all American workers[6].

Bringing us forward to the past decade, we now have an older, retiring Defense Industry workforce. Also a workforce that is in dire need of replenishment, as the Defense Industry is rebounding and growing quickly again due to an increasingly unstable world, as demonstrated by the growth of the US defense budget from $362 billion in 2000 to $614 billion in 2014[7]. You might think that we have this coming-of-age millennial generation ready to fill these opening positions and even grow the ranks, but no one told the Defense Industry PR department, because the image of the Defense Industry flies in the face of the millennial generation's occupational and lifestyle preferences, steering millennials clear of Defense Industry jobs.

First, let's keep in mind the millennial generation is the highest indebted generation coming out of college. In the US "seven in 10 (68%) college seniors who graduated from public and private nonprofit colleges in 2015 had student loan debt…These borrowers owed an average of $30,100."[8] Therefore when millennials exit college they need to make cash ASAP to start paying down their debt, and they quickly learn that the federal workforce can pay up to 34 percent less than their counterparts in industry[9]. That's not bad news for the private sector of the Defense Industry, but it certainly doesn't help backfilling the silver tsunami in the federal workforce. It also doesn't help that as millennials open their first student loan bill, whomever can cut a paycheck the fastest is the best employment option. Yet few sectors are as slow to hire as the Defense Industry, especially the federal component. From my experience, the timelines for the bureaucratic paper processing and mandatory background checks in the Defense Industry can result in comically long delays in job offers. The best candidates will typically have forgotten they even interviewed, already accepted other offers, paid off that loan, escaped their post-college residence (parent's basement), and popped out a few kids by the time the federal HR department gets back to them with that job offer. Additionally, to state the obvious from a human resource perspective, "the longer it takes to hire a good candidate, the less chance you have of actually hiring them. If it is taking 90 to 120 days to get through the process, you end up stuck with the people who are left over versus the people you want."[10]

Unfortunately, this need for quick offers also reveals a behavior that is very disruptive to the Defense Industry, and that is a mobile and fluid workforce, which millennials actually prefer. Can you guess the average tenure of millennial employees? Two years. The average tenure for Gen X employees is five years, and seven years for baby boomers[11]. Millennials are thus known as being "job-hoppers"[12]. Power to them, but in an industry that requires lengthy security vetting and decade-long programs, this fluidity can result in a lack of continuum and stability. It's like turning over the construction of your house every few weeks to a different contractor who has to take the time to read the blueprints, understand what has been done and what hasn't, and what the long-gone architect meant to achieve with the plans. And while your house might take a year or two, defense programs can last decades; thirty to fifty years is not uncommon for aircraft, submarines, and ship programs. This turnover can result in massive inefficiencies, which can lead to program overruns.

This mid-stream involvement is not only tough on the product; it is tough on the people. Millennials are "accustomed to instant gratification."[13] Millennials also prefer recognition for their hard work and accomplishments[14]. Tasks on defense projects are not tasks that manifest in near-term results and accomplishments when you're on the job for only two years, and the program spans multiple decades. Frankly, if a millennial were to become a Defense Industry "lifer," they might be lucky to work on three to five major programs in their entire career, so gratification is less instant, and more on the cadence of the Olympics or World Cup.

The Defense Industry's ties to the military also dictate many of the Defense Industry geographical bases. The problem being that there is little overlap between those bases and millennial hotspots. First, we take Richard Florida's "creative class," his posited class of creative professionals who drive the economy by the creation of new ideas, technology, and creative content, and the cities associated with it. These creative class cities, which are popular with millennials, include Cupertino, CA; Pala Alto, CA; McLean, VA; Bethesda, MD; Brookline, MA; Potomac, MD; Cambridge, MA; Newton, MA; Arlington, VA; and North Bethesda, MD[15]. Next, add what *Time* magazine lists as the top 5 millennial cities: Seattle, Tucson, Columbus, Atlanta, and Austin. Now we'll compare those fifteen locations with reported defense heavy communities. One report lists the top ten defense job locations as, in order, Washington DC, Orlando, Tampa Bay, Huntsville, Colorado Springs, Tucson, Phoenix, San Diego, and Los Angeles[16]. Around the edges folks will also include Austin, Detroit, Silicon Valley, etc.[17]. As we can see, few regions that millennials have colonized have significant Defense Industry presence. The Washington DC area understandably has some overlap; perhaps Austin comes the closest on the millennial-defense Venn diagram center. Certainly, OG hipsterdoms like Portland, Nashville, and Brooklyn have little to no defense sector presence. Perhaps what is most notable is that the most "techy" part of America, Silicon Valley, barely makes the list. Outside of not-so-successful efforts by the Defense Industry to infiltrate Silicon Valley that we'll discuss later, there's a rather light presence of the Defense Industry in the Bay Area. So to enter the Defense Industry, most millennials will have to leave their hot-spots behind for more social networking, because they'll be facing a sharp decrease in per capita brewpubs and coffee shops in popular Defense Industry locales.

When a millennial does takes that opportunity to work in the Defense Industry, they'll come to work and experience two things that are perhaps the most contradictory to their generation beyond delayed gratification and lack of coffee shops, and those are bureaucracy and cybersecurity. "Patience is not a virtue of the young, and Millennials dislike bureaucracy even more than the Baby Boomers and Generation X. Waiting for the gears to grind along isn't their strong suit (especially when there's already an app that can do it faster)[18]." Yet this is an industry where a study showed that almost a quarter of the $600 billion budget was dedicated to bureaucratic operations[19]. That's a lot of bureaucratic grinding, and from experience, it is maddening. Furthermore, perhaps the largest affront to millennial sensibilities will be their limited access to IT in the Defense Industry. Cybersecurity is of the utmost concern in the Defense Industry, in the private and federal sectors alike, which typically results in a limited access to current technologies. Most defense workplaces significantly limit the use of Bluetooth, Wi-Fi, and apps; significantly limit administrative rights on workstations; and use operating systems that typically lag multiple releases. This tends to lead to employees taking matters into their own hands, and overall "younger app users are more likely to go rogue than their more senior colleagues… More than one in three Millennials, and almost as many Gen Xers, work remotely using apps (most often Dropbox or Google Docs) without first getting the IT department's OK. That compares with 22% of Boomers. Some Millennials also take a devil-may-care approach to passwords. About 30% display their app passwords on Post-It notes or somewhere else in plain sight at work, compared to 20% of Gen Xers and just 11% of Boomers[20]." But in an era of increasing cybersecurity, the Defense Industry is frequently audited by government security oversight agencies, and improper actions are actionable offenses that can end up as permanent black marks in a personnel file, if not result in being fired.

We can't afford to have anyone fired from the Defense Industry. We already have Gen Xers and baby boomers leaving the Defense Industry en masse, and on top of those vacancies we have increasing new vacancies being created due to increased global threats. Yet with all of these job openings, millennials might swipe left on the Defense Industry for all the reasons we stated…or will they? In fact, to their credit, more so than previous generations, millennials want a job with purpose[21]. And if you want a job with a purpose, what would provide more fulfillment than assuring our national security? The Defense Industry impacts every citizen through its role in the security of our freedoms, and it reaches to every corner of the globe (and space!), and so few jobs provide as much potential for un-paralleled experiences and fulfillment as the Defense Industry. So keep tracking with me millennials, you'll like where this is going.

Chapter:3
Dear Millennials, Winter is Coming

"I hope I'm wrong, but there's a war coming"
-General Robert Nuller,
US Marine Corps Commandant[22]

Since the millennium, it seems as if winter was coming, per the turn of phrase from Game of Thrones. A new episode of world history began, the peace rave that was the 1990s and post-Cold War ended, and the pangs of new world order changes started to nip at our forehead like a headache. In fact, in January 2018, the "Doomsday Clock," a product of the *Bulletin of Atomic Scientists* that is a symbol of the likelihood of a man-made global catastrophe such as nuclear war, moved to two minutes before midnight, the closest it has been since 1953[23].

For the edification of millennials that haven't considered how we plan our military strategy, our allied national security specialists primarily focus on five major threats: China, Russia, Iran, North Korea, and Terrorists, the last an umbrella for groups spread across the globe that have militarized for a variety of vendettas against our allied nations. International terrorism dominated the early 21st century headlines and still requires the ever-presence of troops in locations around the globe, but the increasing maturity and capabilities of our near-peers China and Russia, and the continued rhetoric of smaller but capable countries like Iran and North Korea, have heightened the "pucker factor" to the next level over the past decade. Additionally, it should be no surprise to say that today, as compared to last year, five years ago, or ten years ago, there is more instability across the globe and more unpredictable threats to the national security of our allied nations. While harder to categorize and plan for, this unrest requires constant monitoring in the event it acts as a catalyst to existing feuds.

Allied nations, their national security specialists, and their defense industries, led by Xers and boomers, have been maintaining a constant vigilance against these threats. But as time passes, it is approaching the moment for the millennials to take their turn at the wall. The following scouting report summarizes what to expect and why.

CHINA

While the world is continuously distracted by events in the Middle East or Russian rhetoric, China utilizes that distractive top cover to continue to grow and grow and grow outside the spotlight. The nation of 1.3 billion people is now the nearest military competitor to the United States in terms of sheer defense funding spent, if not yet capability. China is in for the long game, a commitment to develop their strength gradually to be a great power, at most engaging in defensive measures and low intensity coercion when necessary. With the guaranteed reign of the Chinese Communist party for decades, the execution of this comprehensive and focused plan to utilize the diplomatic, military, and economic forces of China to achieve their objectives is in place with an eye towards 2049, the 100th anniversary of the People's Republic of China.

Obviously a nation with as much technical and manufacturing potential as China is capable of producing the full array of defense technologies to compete with the best powers of the world. Hundreds of millions of iPhones will ultimately be the least impressive footnote of their efforts when history describes China in fifty years. On the other end of technological complexity, China maintains a small but effective nuclear delivery capability to act as a deterrent to any attack, and has invested heavily in anti-ballistic missile capabilities to defend against potential allied missile attack. But per China's calm demeanor, China has declared a "no first use" policy with regards to nuclear strike[24]. On land, China has the world's largest army by the numbers, which they're drawing down to transform into a leaner, more professional fighting force, and in the process, shift those resources away from land to project outward to sea and air to expand their power[25]. At sea, China maintains high numbers of less-than-capable surface ships and prototype aircraft carriers (as compared to allied nations), manages the early stages of a competent nuclear submarine force, and maintains hundreds of unofficial naval vessels (coast guard, friendly fishing vessels) to project power and presence into their neighboring coastal seas, but couldn't muster much disruption far from mainland China. In the air, China continues to field fighters, bombers, and UAVs that lag only a generation or half-a-generation behind allied aircraft. But most importantly, China invests heavily in emergent warfare tactics discussed in the following chapters, intuiting that conventional force-on-force conflict with a group of allied nations doesn't have a high chance of success, but fighting in new domains of warfare where playing fields are equal provides significant tactical advantages.

That said; traditional power and influence still stems from projection from your own territory. I recently found a 1993 issue of an *Islands* magazine that gave this hot travel tip: "And the beautiful Spratly Islands in the South China Sea could be ideal vacation sites – except that they are claimed simultaneously by Malaysia, Vietnam, China, Taiwan, the Philippines, and Brunei[26]." I'm sure a travel writer had nice things to say about Bikini Atoll in the 1930s too, but at least this writer recognized the political sensitivity with the islands, yet probably couldn't have predicted that twenty-five years later the Spratly Islands would no longer be contested, but be unilaterally claimed, occupied by

China, artificially expanded in a massive construction project, and militarized. China's land grabs in the South China Sea this century are one of the most alarming national acts on the planet, by ignoring the claims of at least eight other countries. These military bases will soon provide China uncontested access to massive energy resources in the South China Sea and oversight of over 50% of the world's merchant shipping through the Strait of Malacca. But on another island northeast of the Spratlys in the South China Sea, China has not made as much progress as they wish. Taiwan, an island where nationalists escaped the ruling communists and setup their own government, has evolved into one of the strongest Asian democracies. The US recognizes Taiwan as a sovereign nation and has by treaty agreed to defend it if attacked by China, and to that end, supplies arms to the island. Unfortunately, it does so in the face of China's One-China policy, as Taiwan is part of China in their beliefs, creating a constant tension between two of the world's great powers.

The question then becomes how does the world's relationship with China go off the rails? It could be tied to Taiwan, but it might be tied to money. First, China's economic growth is entering its slowest period in the last generation[27]. That's dangerous for a few reasons, because sometimes a country may more willingly be drawn into a conflict knowing it can't afford a war after the downturn, or a country may use conflict during times of economic stagnation for the purposes of distracting the populace and to rally them, as theorized by some[28]. Yet, it's hard to say if history has ever provided a case of two groups (China and allied democratic nations) so economically dependent upon the other going to war[29], though similar theories were cast about in WWII before it erupted. But unlike the European countries going to war with each other in the World Wars, one difference between China and the world's other powerful nations is distance. The great blue buffer between allied nations and China helps, and that blue backyard is primarily under allied influence for now. As author James Michener said best, "Millions upon millions of years ago, when the continents were already formed and the principal features of the earth had been decided, there existed, then as now, one aspect of the world that dwarfed all others. It was a mighty ocean, resting uneasily to the east of the largest continent, a restless ever-changing, gigantic body of water that would later be described as pacific." If you still have a globe for some reason in 2019, take a look at the size of the Pacific. Google Maps will suffice, but seeing the size of Pacific on a globe really puts its enormity in context. Mostly forgotten by pop culture after WWII and Vietnam, this huge expanse and the thousands of small islands that stretch across it will become a much more common image on our news sites. Its size will help absorb and delay the coming awkwardness of dealing with China's growth, but the US has already begun a proactive strategic "pivot" from the Atlantic to this massive body of water. The intent is to rebalance the US force structure to address the vast Pacific theater, which is envisioned to be the primary focus of national security policy in the 21ˢᵗ century. This pivot means not only more forces (60%) to the Pacific than Atlantic (40%), but includes multiple economic and political initiatives that some view as tools for China's containment of their military power, but not yet their iPhones.

RUSSIA

While it is certainly not time yet to return to the duck and cover drills from grade school in preparation for Russian nukes falling from the sky, the relationship and rhetoric between the US and Russia is definitely not in a good place. In 2019, the US and Russia stepped closer to the brink by abandoning nuclear missile treaties[30]. After two decades of hibernation, the Russian bear is back. While jailing Pussy Riot, riding horseback shirtless, and providing us many of our favorite memes and countless "Saturday Night Live" skits, Russian leader Vladimir Putin has also been able to shake off the rust from their Soviet military apparatus and funneled significant investment in its modernization, resulting in a small, but extremely proud, proficient, and capable armed forces.

The long history of manufacturing infrastructure and military decision retained from the USSR with regards to fleet force composition, logistics, and the Defense Industry in many ways help make it the nearer peer competitor to the United States than China. Russia maintains the largest stockpile of nuclear weapons on the globe and many ways to deliver them, be it from fixed silos, mobile launchers, or submarines[31]. From land, their army is sizable and supported by the world's largest number of tanks and artillery, though limited to Soviet era technology[32]. The world's 3rd largest air force resides in Russia, and technically, has always lagged just behind the US in terms of capability, though that gap is now closing with China who threatens to surpass Russian capabilities. At sea, Russia has written off a significant portion of the Soviet fleet, slowly rebuilding into a small and capable force, with submarines specifically as their capital ship to support multiple regional objectives[33]. Russians have always excelled in submarine technology, exporting their quiet, diesel Kilo class to many nations during their economic downturn, but are now building and deploying new submarines for their own navy at Cold War levels[34]. In all domains, allied nations must always be sensitive to the latent, sleeping infrastructure of companies and knowledge pools left after the Soviet era, which have been propped up heavily in the interim by foreign export sales, but Russian economic stabilization over the past decade has turned their focus back to the motherland. Lastly, much like China, Russia has invested heavily in cyber capabilities, and while not purely state-sponsored, Russia is typically referenced as the top cyber threat to America, and like America, is standing up their own cyber command[35].

So all told, the Russian military might is back. When we say their military is back, I mean it's at our doorstep, more so than any other nation. Russian flight incursions into Baltic airspace intercepted by NATO have occurred over 100 times per year in the past three years[36] and in UK airspace at least a dozen times[37]. Likewise, flexing their muscles in their Pacific backyard, Russia has been conducting operations near Japan, Hawaii, and Guam, and while "Russia stopped conducting regular bomber patrols in the 1990s and early 2000s, (it) has increased its patrol activities in the Pacific Ocean following the Ukraine crisis in 2014 amid the resulting isolation from the West… Between April and December 2016, the JASDF scrambled its warplanes … 231 times in reaction to the Russian Aerospace Forces[38]." From the sea, Russia deployed its largest naval task force since the Cold War in October 2016, and their survey ships have deployed near Hawaii and the US east coast multiple times in the past few years[39,40]. While this is commonplace frankly for allied operations across the globe, and commonplace in the Cold War, the ability for news outlets to live stream live imagery of Russian naval ships and bombers lurking over the horizon off our coast is more powerful in our new modern way.

Ultimately, far from being just reactive and probing the defense of potential adversaries, Russia has used its new forces to expand its influence and capabilities. On par with China's island land grab, Russia annexed Crimea in 2014, and has since also committed military support to the government in Syria and expanded bases in the Arctic. These actions certainly make their rhetoric more than must-listen radio, but it must be wondered if Russia is a paper lion, ready to repeat the mistakes of the past. The Russian economy is significantly hampered by corruption and economic volatility caused by Western sanctions, so can it continue such a pace of modernization? And if not, would anyone in Russia prevent it, or would its corrupt political and economic leaders look out for their own first at the cost of another collapse? These are the questions that will play out during the upcoming years.

IRAN

Iran continues to contest allied national interests in the Middle East. If anything, they act as the leader of a gang of organizations that are allied against the United States, Saudi Arabia, and Israel. Their trump card is the continued pursuit towards nuclear and theater intercontinental missiles (up to 3500km) that can reach our allies in the Middle East and even Europe. They are certainly the most capable navy in the Persian Gulf with surface craft and submarines, projecting power out through the Straits of Hormuz and the Gulf of Oman, and even out to the Arabian Sea, the Gulf of Aden, the Red Sea, and the Caspian Sea. In the air, Iran struggles, leaning on fairly antique fighter craft that were built prior to the Shah being overthrown in 1979[41]. But that is not to say it'd be possible to just fly into Iran. Iran has a very capable air defense system for anti-access/area denial, consisting of missiles and radars and potential satellite jamming and interference capability. Furthermore, they've also invested heavily in unmanned air vehicles, reverse-engineering US drones that they've allegedly captured. On the ground, they have a sufficient sized infantry, but are not well-equipped or sized for any significant battle beyond their borders. Thus, ultimately Iran relies on asymmetric equipment and abilities like missiles, mining, and unmanned vehicles to deter any attack or interrupt commercial interests in the Persian Gulf, such as the mining of oil ships in 2019.

What Iran does better than most against the US is to pick at open wounds, providing support to Syria, Yemen, and Hezbollah. They have also developed significant ties with Russia and China for arms purchases, providing reciprocal support in the form of allowing ship visits and refurbishment to facilitate distant naval deployments of those countries, which helps contest the influence of the US Navy in the Middle East. Yet while run by a strong theocratic government, dictated by the religious conservatives, it is commonly reported that the majority of the Iranian population does not hate the US, and there is a large and mobile Iranian population in the US that frequently travels back and forth between the US and Iran, exchanging information and culture. What it means to have the threat that is Iran with a population more inclined to allied alliances than its counterparts is unknown at this point, but certainly it plays a role in the future of this relationship under millennial watch.

NORTH KOREA

The crazy uncle of state actors, constantly making inappropriate comments during holiday dinner, only to be found the next morning passed out drunk on the backyard trampoline. North Korea is perhaps the most surreal of our enemies, as demonstrated by its humiliating fictionalization in such movies as *The Interview* and *Team America*, and then reinforced by real life events like Dennis Rodman's interaction with Kim Jong-un as covered by Vice TV Network, or the murder of Kim Jong-un's half-brother, leaving the general impression of North Korea as a nation with a slight chemical imbalance. Yet that crazy imbalanced uncle has nuclear weapons with intercontinental ballistic missiles to carry them, and dozens of short range submarines, and an immense cult of personality that might be strong enough to convince their army of over 8.9 million to trample straight over South Korea and the 30,000 US troops stationed there while they're at it[42]. Not that they have the logistics to get any farther, but they have developed a burgeoning cyber warfare capability, and those missiles we see in the news every few weeks could certainly cause significant damage to South Korea, Japan, and just maybe Guam and Hawaii. As intended, international sanctions have continued to hamper their technological advancements, but North Korea's inability to acknowledge the sensible logic of mutually assured destruction makes them one of the most dangerous wild cards in recent history.

TERRORISM

In the summer of 2001, I took a graduate class on terrorism that ended a few weeks before 9/11. When I look back at that class syllabus, it was indicative of every failure we admitted post-9/11, spending almost 80% of the class on domestic terrorists inspired by racism, nationalism, or divisions between western religions, like the Basques or the Irish Republican Army. It's astounding how naïve we were at the time. Since then the War on Terror has been estimated by some to cost over $2 trillion[43]. We continue to fight ISIL, Al Qaida, and legions of homegrown violent extremists throughout Syria, Iraq, Pakistan, Afghanistan, Yemen, Somalia, North Africa, and South Asia. These brands of terrorists notably fall into extremist religion ideology and not the political reasoning we've associated with terrorism in the past. It is difficult to "win," as the ideology brings extreme devotion to the cause, and as it is with de-centralized, non-state actors, there is no formal surrender, at times no formal leadership, and so these terrorists will continue to threaten our nations for many years to come under millennial watch.

Terrorists primarily utilize asymmetric warfare, attacking so-called "soft" targets like innocent civilians and critical infrastructure, versus military and executive targets, as demonstrated in the attacks throughout Britain and France in 2016 and 2017. Making use of a globalized information technology, they will use modern media and the internet to recruit and spread propaganda. With respect to our defensive response, the War on Terror has less to do with mass mobilization and conventional warfare and more to do with surgical use of Special Forces, intelligence collection, and cyber warfare. These methods help allied forces to conduct surgical strikes against key nodes, win the hearts and minds of the populace that is typically recruited by terrorists, interrupt funding mechanisms for terrorists, and assist struggling nations, as terrorists prey on the disaffected and disillusioned.

GLOBAL INSTABILITY

Lastly, instability is an increasing challenge across the globe that complicates future projections. It's out of this instability that a minor event could occur that might snowball into a conflict that was never predicted. Per the 2015 Global Peace Index report, "the world is less peaceful today than it was in 2008. The indicators that have deteriorated the most are the number of refugees and IDPs (Internally Displaced Peoples), the number of deaths from internal conflict and the impact of terrorism. Last year alone it is estimated that 20,000 people were killed in terrorist attacks up from an average of 2,000 a year only 10 years ago [44]." If it feels as if the world is seeing more protests, it's because it's true. "Major citizen protests are multiplying. Just in 2015, significant protests erupted or continued in Armenia, Azerbaijan, Bosnia, Brazil, Burundi, the Democratic Republic of the Congo, Guatemala, Iraq, Japan, Lebanon, Macedonia, Malaysia, Moldova, and Venezuela. The list of countries hit by major protests since 2010 is remarkably long and diverse. It includes more than 60 states that span every region of the world[45]." Even locally, those in the US certainly feel a larger divisiveness than in the past, as exposed by the gap between the left and the right after the 2016 presidential election. To prove the reality of that divide, it's reported that "in 1997, there were 164 congressional districts that performed somewhere between five points more Democratic and five points more Republican than the nation as a whole. In the 2016 election there were only 72 of those seats"[46]. Districts are increasingly non-competitive red or blue, inherently preventing discussion and debate of ideas. And on top of all this, "risks to the global growth outlook are two sided but are assessed to be skewed to the downside" by the IMF[47]. A downturn in the global economy and the impacts of a Brexit could even further destabilize world order and increase stress amongst nations and their peoples.

WINTER

Near peer competitors, aggressive or failed countries, terrorist attacks, massive refugee movements, protests, political divisiveness, and financial downturns: any or all could result in events that might threaten national security and will need to be monitored with a close eye. The national security environment is fluid, constantly shifting every day, and this summary is likely to be overcome by events even before it's printed. But at minimum as an exercise, this chapter helped frame the wide-array and perhaps unpredictable types of external challenges millennials will be facing in trying to secure our national security. Architecting a Defense Industry that can address all of these threats will certainly require a high degree of art, engineering, and business acumen.

To that end, a slew of preparations have been made in advance of the approaching "winter" by previous generations. In the United States for example, the preparation originates from the Department of Defense led Joint Capabilities Integration and Development System (JCIDS) system, which ultimately shapes the US defense industrial complex. The JCIDS process identifies, assesses, validates, and prioritizes military capabilities, and from that builds the ultimate shopping list required to engage threats via the hundreds of war plans and contingency plans that have been drawn up over the decades. Similar processes occur in our allied nations, and allied nations work with each other on assessing and prioritizing threats, and even shaping forces based on knowing how and where our allies can compensate for each other's weaknesses. These processes will help millennials prepare for the oncoming winter, starting the production lines of the items they'll need for the next battle. So it'd benefit millennials to engage this process soon, as the JCIDS decisions made now will impact the success of national security decades down the road.

Chapter:4
Lasers, Rail Guns, Cyber, Oh My

"The more you sweat in times of peace,
the less you bleed in times of war"

-Proverb

Regardless of how millennials fix the Defense Industry, one thing we know for sure is future Defense Industry production lines will be shaped around some mind-blowing, goals AF technologies. Millennials will be ushering onto the battlefield technologies that were once the realm of Futurists and comic book writers of the twentieth century. Playing directly to a millennial's strength, there will be a significant crafting of these technologies to achieve valuable effects on future warfare operations as these technologies transition from the laboratories to the battlefield for the first time. But it won't be easy; it'll require long nights and weekends to get these technologies to work, but it'll be worth it for the defense of our freedoms when they do. Certainly not comprehensive, but a few of these exciting technologies are discussed hereinafter.

ROBOTS

Millennials missed out on the dawn of the Bill Gates and Steve Jobs operating systems (OS) that now dominate the world. That historical level of computer contribution to humanity has sailed, so what will millennials be remembered for? What the world needs is an operating system for a machine brain, not an operating system that runs software to help the human brain. For example, the robotic operating system (ROS) is the popular middleware for many robotics companies building unmanned and autonomous vehicles. Many robotics companies basing their products in ROS are disrupting the Defense Industry; larger defense companies are gobbling many up as they recognize the future of robotics in warfare.

But defense and autonomous robotics certainly had an inauspicious start. One of the initial defense-funded highlights, and amusing failures, of unmanned technology consisted of the DARPA Grand Challenges back in 2004, with unmanned vehicles getting no farther than 7 miles into a 150-mile course! But since then, a series of DARPA contests, including an urban challenge where vehicles were successful in unmanned navigation, have certainly been a contribution to the unmanned car industry we see blossoming today.

Coupled with almost ubiquitous use of unmanned air vehicles in battle zones such as the Predator, Reaper, Firescout, or the sophisticated and mysterious unmanned X-37 military space plane that has made multiple space deployments and successful landings back on Earth, or the burgeoning hobby drone industry, there is a significant base of experience and knowledge regarding unplanned platforms in industry. And not to be left out, the undersea elements have also begun to re-architecture their forces, operational support, and research divisions to include extensive unmanned operations. The MK18 Kingfish and Swordfish programs, for example, have conducted thousands of operations with Hydroid REMUS vehicles, which were born from oceanographic research and shipwreck hunting and are now pushing deeper into the battle space. There are now unmanned vehicles operating in every domain on the battlefield.

Now will all these robots stop fighting each other, team together, and overthrow our human society? Not yet, they'll probably save that for the next generation (sorry Generation Z), but robots will proliferate in-use and be a component of almost every aspect of national security. Robotics introduction was initially in manufacturing; bringing out big strong dumb robots on the construction floor, where they truly did take many human jobs and made no friends[48]. But the robot market will shift considerably in the next few years to robots that will be "personal assistants, delivery vehicles, surgical assistants, exoskeletons, autonomous vehicles, and unmanned aerial vehicles (UAVs), among many others [49]."

The trends in unmanned robotics will primarily be in the realms of 1) pushing manned systems into hybrid manned or unmanned versions, 2) transitioning those already unmanned systems from remote to autonomous, and 3) pushing those same unmanned systems into a society of their own such that they work together towards an objective, or perhaps augmenting human-operated platforms with a number of unmanned or autonomous vehicles to support them, meaning "what used to be solo systems are going to have to be teams to be relevant in the near future[50]."

This state of affairs with robots will be enabled not just by ROS and emerging unmanned system architectures, but also by further advances in deep learning and machine learning. Artificial intelligence will allow robots to increase battlefield autonomy to near full-level automation. That being said, when UN law and the Geneva Convention governs the battlefield—insert a lawyer joke—law is one field that robots are not yet threatening with their intelligence, but quite the opposite. It will require a significant increase in lawyers to help define how law and ethics applies to these unmanned systems, on or off the battlefield. The unmanned battlefield will not be just an engineering feat, but one for humanity's relationship with technology.

CYBER, ELECTROMAGNETIC, AND SPACE WARFARE

New cyber and electromagnetic battlefields will result in whole new ecosystems of defense infrastructure. If defense budgets remain flat in the future, these domains will chip away funding from existing forces, creating the natural tension between old world warfighters that can't grasp combat they can't see and touch, and new world millennial warfighters who see the gaping vulnerability in our cyber and electromagnetic defenses and capabilities. The US Cyber Command is an example of these new, rapidly growing organizations consuming personnel from traditional organizations. Incubated in the National Security Agency, and bringing together elements from all the US service components (Army, Navy, Air Force), Cyber Command is working hard just to get the staffing increased to its planned 6,000 person, 133 team force[51]. As a demonstration of how hot the need is for personnel, the US Navy is even looking to allow for lateral entry into the navy (e.g. skipping all those early rungs) for cyber professionals, a maneuver they typically reserve only for lawyers and doctors[52]. What is the reason for such demand? Millennials, and all of us for that matter, must realize we're already at war in Cyber. While many boomers and older generations were used to the proxy wars of the Cold War in Asia and Latin America, Cyber is the modern day equivalent, but more insidious in that it helps mask the attackers and their intent; it is a lawless and anonymous wild, wild west. It is also cheap and asymmetric; international law governing such attacks provides little strength of response, at least for now, until countries can stand up the appropriate commands.

Allied nations have not put enough forethought into the impact of transferring much of their power into digital ones and zeros, hidden behind frequently poor levels of cyber security. Cyber warfare is about twenty years old, but we've seen examples of how it can run the gamut from significant breaches in security of companies (like 1 billion Yahoo accounts compromised), financial institutions (the hacks of Mastercard, Visa, and Target) or the government (the US Office of Personnel Management breach, which resulted in the release of up to 18 million US federal employees and their clearances). Of all vulnerabilities, we've yet to see a major attack on a power grid or other infrastructure, but many warn that it is inevitable and near[53]. These previous examples were surgical strikes or misfit hackers, not the full collective brunt of national resources brought to bear in a declared cyberwar. What will be much different in societal response to the breakout of cyberwar is that traditional battlefields did not spillover into the everyday life of civilians, but cyber warfare will; it will impact their identities, their bank accounts, their infrastructure, it will in a sense be one of the most democratizing forms of warfare that will place everyone, unwillingly, on the front lines. Everything that is internet connected is a vulnerable target, from cars, to phones, to photos, to your money, to your TV, to your home security. They would all be targets in a cyberwar. As that begins to sink in, most nations are now standing up cyber commands staffed by teams of cyber soldiers to fight these battles, and the early stages of these commands will come under the millennial watch, before they inevitably become services alongside these nations' armies, navies, and air forces.

Electromagnetic Maneuver Warfare (EMW) on the other hand is still in its infancy, but one thing is certain, that it will impact doctrine and system design for the next century. Imagine the coffee shop as the battlefield, and the shop owner as the adversary. When war starts, that shop owner will turn off the WiFi, and if they're especially dastardly, will install a cellular signal blocker. But you still want to sit there and finish your cold pressed coffee and edit that Instagram picture, so your friend outside on the patio beyond the cell jammer gets a signal and creates a new local network to bypass the shop owner's offensive. That's an example of EMW. "The electromagnetic spectrum is an essential - and invisible - part of modern life [military and civilian]. Our military forces use wireless computer networks to coordinate operations and order supplies, use radars and sensors to locate each other and the enemy, and use electronic jammers to blind enemy radars or disrupt their communications," states Jonathan Greenert, former Chief of Naval Operations. "With wireless routers or satellites part of almost every computer network, cyberspace, and the electromagnetic spectrum now form one continuous environment[54]."

Up until now in history we have primarily operated in the electromagnetic spectrum with brute force and with impunity, meaning we communicate on specific radio frequencies, we emit radar at specific frequencies, and we do it at some level of signal power, with full disregard to the environment or what signals are already out there. It'd be like looking across an actual battlefield and noting the terrain and its impact on an advancing army. Along one side of the field might be a steep hill that'd need to be circumvented, but in the electromagnetic spectrum, we'd behave as if it wasn't even there, where in reality the weather or the enemy, through jamming, might be preventing transmission (e.g. the hill). This perception of clear and level battlefields in the electromagnetic spectrum is over, we won't be able to conduct our operations effectively if we don't recognize that, and we certainly won't understand what our adversaries are doing if we don't concede that. In this new concept, our allied "aircraft, surface ships, and even submarines will have sensors to pick up electromagnetic emissions in their environment and data links to transmit what they find," and then our allied forces will use this electromagnetic environment to behave like an army; they'll hide in it, they'll exploit gaps in it, they'll attack in it[55]. This is how submarines have operated in the acoustic spectrum for decades, managing their signals such that they were invisible "holes in the ocean," maneuvering through acoustic frequencies to their tactical advantage. If our allied nations don't do a better job at managing the electromagnetic spectrum, our adversaries will exploit that unprotected path into our defense systems, putting all our data, information, and knowledge at risk.

While the actions of cyber and electromagnetic warfare soldiers are conducted on Earth, many of the paths into communication networks of interest rely heavily on the telecommunication paths from Earth to space and back into radios, World Wide Web, and other IT devices, and thus present significant targets of interest for adversaries. If warfare were to enter space, it'd be a tipping point in international relations on par with America's decision to drop the first atomic bomb. First, that would mean decades of attempts at laws related to keeping space peaceful have ended, and second, the entire planet could be impacted by the loss of use of specific satellites or satellite constellations, especially if satellites were physically destroyed and sent space debris in orbit that could then damage other satellites (and astronauts). For these reasons, space has been treated similar to the Earth's poles. While accessible to militaries, space is not owned by any nations and has been the subject of many treaties, but space and the poles both represent such significant high ground in the event of war that they will be difficult to protect in that circumstance[56]. Space is so highly integrated with the operations of a nation in war, from communications satellites, to GPS satellites, to intelligence satellites, that it is hard to blame an adversary from wanting to pierce the veil of space ethos and actively disable satellites, as that would significantly inhibit any country's ability to execute war against them. To this end, many potential adversaries have demonstrated anti-space capabilities. China "tested an anti-satellite missile and destroyed one of their weather satellites, a move criticized because of the debris field created in space[57]". Likewise, "Russia has sent micro-satellites into space and covertly maneuvered a small spacecraft close to commercial satellites. Experts believe the small satellites could be used for a kamikaze-type mission to ram another satellite or to snoop on it for data collection or jamming to interfere with its capabilities." One can only hope that if an offensive on space assets were to happen, nations would focus on disabling satellites by cyber-attack or by jamming the signals, rather than kinetic options. Regardless, it is quite likely millennials will be the first generation to manage the fallout from the end of space innocence.

BIOTECHNOLOGY

Professional athletes looking to keep in good standing need not apply, because the future of soldiers will include cocktails of supplements to improve endurance, strength, focus, and intelligence. Attempts at these advancements include "DARPA's Peak Soldier Performance Program, which in 2004 sought a biochemical approach that would allow a soldier to operate in theatre for up to five days without requiring sustenance[58]." In more recent times, US "Special Operations Command wanted to create super-soldiers through pushing the limits of human performance, and is looking to nutritional supplements and even performance enhancing drugs as options[59]." As the human body has its very real operational limits, the second aspect of biotechnology is the augmentation of the human body with technology to improve performance. This result is perhaps acutely appropriate for the millennial generation that has defied labels of sexuality and sex, because the potential goal here is to tear down the next wall, that of human, for the "trans-human" soldier, the coupling of technology to the biological human being with a dual-focus on how to build an optimum human for the battlefield and/or how to repair that human being after battle, yet retain a decent quality of life. Regarding the former, the best analogy might be Iron Man. The Defense Industry is trying to build Iron Man suits for soldiers. The US Special Operation's Command built a prototype in 2018. It doesn't fly and shoot lasers like Iron Man, but it is meant to be a lightweight protective armor equipped with sensors[60]. All of this is apparent from a business perspective too, as "biotechnology is one of the world's fastest growing commercial sectors. Since 1992, the number of biotechnology companies in the United States alone has tripled... Many of biotechnology's benefits are dual-use, increasing the possibility that knowledge, skills, and equipment could be adapted for use as biological weapons. As the global biotechnology industry expands, the U.S. government should therefore increase its capacity to exploit biotech advances for national security[61]."

SUPERSONIC TECHNOLOGIES

While bio-enhancement will make our soldiers run faster, up in the sky our planes will be swapping out their jet turbines for the new millennial models to kick it into the next gear. The first century of jet engines, sure, they went fast, they could at times break the sound barrier, which equates to supersonic speeds (greater than Mach 1, the speed of sound), but that's just not fast enough; the world demands Mach 3 and 4! The futurists and idealists say that's not even enough, we need hypersonic (Mach 5+). If we make our jets and weapons go faster, we can get to the target faster and neutralize it before it does more damage, and we can evade our adversary's weapons.

The crazy thing is that these aren't that far away. The engine of low sonic speeds, the turbojet we see strapped onto all our planes as we fly from city to city, is so 20th century. The future is ramjet and scramjet, and defense and commercial entities are racing towards the goal line (though a lot of conspiracy reports will tell you that secret spy planes already use these technologies). One of the higher profile and public experiments includes the recent tests of an international program, the Hypersonic International Flight Research Experiment (HiFIre), in which missile travel achieved Mach 7.5[62]. Additionally, NASA and the Air Force have multiple "X" vehicles, like the X-43 and the X-51, testing scramjet propulsion[63]. These engines can be utilized for planes or missiles. And what does that ultimately mean? It is another reason to be jealous of millennials, because they'll be able to get from New York to London in less than 45 minutes, or LA to Sydney in less than 2 hours.

GRAPHENE

When you're going hypersonic, you'll need materials that are stronger and lighter. And when you win a Nobel

Prize building such materials, you know you've done something right. That's what happened to the scientists at the University of Manchester in the UK for their work in isolating and characterizing graphene, a material that has been theorized about for decades. Not to pump it up too much, but *Nature* magazine puts it well, "it is as stiff as diamond and hundreds of times stronger than steel — yet at the same time is extremely flexible, even stretchable. It conducts electricity faster at room temperature than any other known material, and it can convert light of any wavelength into a current. In the decade since graphene was first isolated, researchers have proposed dozens of potential applications, from faster computer chips and flexible touchscreens to hyper-efficient solar cells and desalination membranes[64]." This stud material also isn't shy; graphene is made from carbon, the fourth most abundant element on Earth. This super material, basically a very thin layer of pure carbon organized in a hexagonal honeycomb lattice, is readily available to any suitor, and can rewrite the lightweight armor game, provide adaptive camouflage, offer new power solutions, and a million other applications on the battlefield if it replaces silicon and the processing game. Of course right now it's a little high maintenance, requiring huge investments to fabricate, but if millennials can make this their generation's plastic or silicon, it will completely change the battlefield.

WEAPONS

So somebody wants to shoot something, perhaps it's the raccoons raiding their garden or their trash... should they go grab their eighteenth century canon? How about a massive, pivoting gun from the decks of a World War II battleship? Or perhaps just a compact modern day handgun they have in the safe? Centuries have passed, and our concept of "shooting" is fairly stagnant. First, ignite that chemical propellant, pushing the bullet forward toward the target, such that the impact either by the kinetic force or explosive in the bullet introduces extensive damage to the receiving end. But two future weapons are going to significantly change that methodology. The first relates to the projection force. The navy has a prototype electromagnetic railgun that shoots projectiles up to 125 miles and destroys armored targets with ammunition at speeds of 4500 mph[65], meaning if you ever played with two magnets and notice how they push away from each other, the railgun does the same; it pushes a projectile out of the gun, just really damn fast. As the Office of Naval Research explains, "with its increased velocity and extended range, the EM Railgun will give Sailors a multi-mission capability, allowing them to conduct precise naval surface fire support or land strikes; ship defense; and surface warfare to deter enemy vessels[66]." No chemicals, safer launch, and more force; that'll teach those raccoons.

Second, LASERS! YES! Think back to your favorite science fiction movie or book (Star Trek, Star Wars, etc, and also the topic of one of the best defense comedies of the

1980s—see Chapter 7: Netflix and Defend). Laser technology is finally in the battlefield…sort of. We're not yet talking soldiers firing holstered laser guns, but hulking kilowatt lasers sitting on navy ship decks, air force planes, or army humvees. For example, the US Navy has deployed the LaWS (laser weapon system), which can disable or destroy targets, depending on how much you turn up the juice. Think of an annoying drone buzzing around the boat, or a small boat making a beeline for a destroyer (like the USS *Cole* terrorist attack), and then—laser sound effect— down goes the problem. Much like the electromagnetic railgun, the best part of laser weapons is that all you need is energy, no ordnance that must be "designed, manufactured, handled, transported and maintained, and takes up storage space[67]." Long as the ship still has power, it has laser weapons; it's not running out of bullets. Lasers and railguns also provide a complementary capability; some scenarios will require projectiles (lasers don't typically have an over the horizon trajectory), and there are moments where lasers will be more appropriate. But it's still early, and under their watch, millenials will need to increase the power on these capabilities while simultaneously shrinking laser technology to proliferate the opportunities and locations to use them.

ENERGY

These electric weapons provide the perfect insight into future warfare. Where first human strength and then chemical processes were once king, electrical forces now threaten to overtake. But how do soldiers, ships, satellites, and planes bring more power, make power, and store power in more efficient means? This will be the future design challenge for millennials due to the increase in electric motors, electric weapons, and increasing computer-processing requirements. As Captain Kirk once said, "Scotty, I need more power." So in light of our electrical power surge, our defense infrastructure needs alternative forms of power generation and storage, renewable energy, etc. In fact, "the Department of Defense (DoD) is the single largest consumer of energy in the United States. As early as 2006, the DoD was spending about $10 billion on mobility fuels[68]." Think about how renewable energy may impact that budget, and what that would free up for other purposes. The list of advantages are long: "(US) military officials argue to Reuters that this shift to renewables isn't really motivated by a desire to save the planet, but to make systems more efficient, safe, and robust. For instance, an Army facility running on renewables would be immune to grid attacks; a hybrid tank doesn't need to stop to refuel as often; and in war zones a solar panel can't explode like a tank of gas[69]." But these energy replacements will only be viable if they can support the increased energy requirements. The Flux Capacitor only required 1.21 "jigowatts" to get going, but our future ships, planes, and tanks may be talking about terawatts and pentawatts once they've loaded every corner of these platforms with computer servers and laser guns. Where older veterans once saw vehicles that were mechanical and pneumatic, millenial platforms will be shells for server farms and electronic devices, and so energy will be the primary constraint to their success.

BACK AT THE LAIR

So obviously the remote, mountainous, island lair of the millennials will be full of new technologies and quantum computers, built for the battlefield revolution of the 21st century and paid with cryptocurrency based on blockchain. These new technologies will be increasingly non-kinetic, increasingly information technology based, increasingly human-out-of-the-loop, and increasingly not in the three dimensional world that humanity has operated in until now. Getting these technologies to work, especially on a battlefield, will keep the lights on the lair at all hours, but these are not the real problems that millennials will face. Scientific problem solving is fun, tinkering is fun, designing and building is fun, experimenting is fun. As most science, technology, engineering, and math (STEM) employees will tell you, the STEM portion of their job is fun, it's all the "other stuff" that is the problem.

Chapter:5
99 Problems

"Some things are beyond my control. For example, this whole controversy about Jay-Z going to Cuba - it's unbelievable. I've got 99 problems and now Jay-Z is one."

-Barack Obama,
President of the United States

BROMSLJUS UR FUNKTION. My car flashed this warning at me every time I turned it on. My car was a Saab, an automotive company no longer, but former producer of great cars born from a Swedish defense company, including a night mode where all the illuminated readouts in the car went dark, except for the speedometer, giving the car a fighter jet cockpit mode straight from its familial defense production lines. But my Saab was under the mistaken impression I had a tail brake light out. This went on for about a month and, unwilling to go to the dealer to get the computer or sensor fixed, I switched the language of my car's display to "Svenska," diminishing the annoyance some as I no longer understood it. That's why my car warned me Bromsljus Ur Funktion for months, a minor issue.

This was happening to me inside an airport parking garage, after returning home to the east coast for a few days from visiting a military base in San Diego, and before heading back out on travel for a few days on a nuclear submarine. The sub was built American and used primarily English warnings when things went wrong, though sometimes the warnings included sounds or flashing lights with better nuance. The trip was intended to test upgraded torpedoes, but many failed to hit their target, including one torpedo I had become particularly close with. My bunk was in the torpedo room, and this particular torpedo was stowed adjacent to my rack. When a lot of visitors are on a submarine (think packed subway car), the unfortunate location for extra bunks and their occupants are in the torpedo room. Each night I'd crook my leg over this torpedo while I slept. Sure it seemed odd at first, but once you've seen how narrow a bed on a submarine is, the opportunity to gain sleeping room reduces one's normal above-sea hesitations and standards. We spent the bulk of our awoken hours trying to diagnose their failures to no avail, a glitch left to analyze back on land.

My car, my torpedo, those weren't engineering problems, just technical challenges that needed to be solved; the laws of physics and disciplined troubleshooting eventually explained them. Technical challenges can be categorized in alarmingly small subset of likely issues, categorized nicely in theories like the TRIZ (theory of inventive problem solving) approach. I'm comfortable knowing nerds will nerd and build the best damn defense widgets integrating capabilities in Chapter 4. Likewise, the threats discussed in Chapter 2 seem like problems, but they are outside of our control; we can't solve them, we can only mitigate them through preparation and planning. Politicians will politic and spies will spy and strategists will strategize and we'll build the best damn diplomatic and war plans. Millennials have 99 problems, but technological feats and national security threats aren't two of them, those are opportunities. The 99 problems millennials will deal with are the processes we've created to acquire defense systems, processes that are rarely optimized for value, rarely optimized for the global security environment du jour, and rarely optimized for maintaining our technological superiority.

Problem number one, say it together, "the bureaucracy," the bane of any defense organization since a caveman went into business whittling spears. The struggle is real. Bureaucracy is the piñata of all that ails the Defense Industry, but it is a necessary evil, and provides necessary checks and balances to ensure that the Defense Industry is appropriately utilizing taxpayer funds. The defense acquisition process of the US and our allied nations produces products that are 1) dangerous, and 2) need extensive vetting to be incorporated into battlefield use. Ask yourself, would you enter a fight with your life or your family's life on the line without knowing whether your weapon works? You wouldn't, and so that weapon needs extensive development and testing. Additionally, defense products are typically expensive (insert fake astonishment), and so intensive management of that funding is part and parcel of the fiduciary duty of the Defense Industry to utilize those taxpayer funds effectively, thereby preventing news-making examples of fraud, waste, and financial abuse, like kickbacks, new cars, strip clubs, and vacations.

However, yes, there is way too much bureaucracy. My colleagues and I at my navy lab had a 20/20/80 hypothesis. We knew 20% of the workforce performed 80% of the work, as the saying goes, but we also believed that the hardworking 20% was limited to performing valuable work during only 20% of their time. So this meant 20% of the workforce working 20% of the time was doing 80% of the job! Talk about room for improvement! If you haven't worked in a massive bureaucracy before you're probably wondering, "what takes up the rest of your time?" Instead of answering that directly, I'll leave it to the professionals, like the comic *Dilbert* or the movie *Office Space* to explain, because they are drawn from reality. But unfortunately, many millennials may actually find themselves in a real life Swan protocol from the TV show "Lost," entering in a sequence of characters in a computer every 108 minutes as their job.

So you'd think that at this point a boss would step in and fix this, push to obtain more efficiency from their employees. But bureaucracy also has unique impacts to the dynamics within the work environment. Most defense organizations, from federal to commercial, utilize a matrix organization, where one has both a project manager and an administrative manager, thus dividing allegiances and priorities, and at times creating a "formless state of confusion where people do not recognize a 'boss' to whom they feel responsible[70]." What compounds and institutionalizes this sense of anarchy is that there is little "hard power." Bureaucracy has made it difficult to terminate people for poor performance, managers are required to collect reams and reams of caught-you-red-handed data to prove an employee is not performing their duties, and in the federal workforce it's near impossible to be fired. Thus managers have primarily "soft power," forcing them to lead employees primarily by appeal and begging, rather than by direction and coercion.

Luckily, some authors such as Daniel Pink, a best-selling author about work behavior, would propose that what motivates us is not financial, but satisfaction of the work, and our human need to prove our self-worth. This bodes well for the Defense Industry, as self-satisfaction should be easy based on the nationalistic motivation to protect our allied nations, overcoming the state of confusion. In my experience this singular aspect is indeed what keeps the best employees in the Defense Industry, regardless of their frustration with the bureaucracy. Remember, at the risk of sounding like a broken record, the millennial workforce strives to work for a purpose. Admittedly, it is also motivating to feel threatened and singled out, and indeed that is the case as our adversaries are constantly working to inflict damage to our nations as discussed in Chapter 3, and that helps drive the Defense Industry along, putting some pep in our engineering step. But that drive to continuously be vigilant, hardworking, and patient in the face of bureaucracy takes grit. It takes a little madness. In the Defense Industry, this madness can be equally important as intelligence, and we'll discuss that more in the next chapter.

Problem two is predicting which of the widgets we've discussed to bring to the fight. The constantly evolving realms of warfare can be seen in the evolving structure of the US Department of Defense, which started as the War Department to run armies in 1789, then spun off the Navy Department to run navies in 1798, finally spinning off the Air Force in 1947. Now the hottest concepts in warfare are Electromagnetic Maneuver Warfare, Cyber Warfare, Information Warfare, and even a renewed Star Wars interest (Space Warfare). Allied nations in most cases have reacted to emerging needs, the World Wars and Gulf Wars in the 20th century being examples that were revolutionized by the introduction of aircraft, submarines, and atomic and precision weapons. In the future, allied nations will need to respond with similar success to new battlefields and threats, successfully prioritizing and estimating their needs through the JCIDs process, and building a Defense Industry that's building the right widgets.

Problem three is one of our most frustrating failures, and that is developing our best ideas, and then advertising this intellectual cash to the world's pickpockets like a naive tourist. We protect our local farmers, as heralded in the proliferation of "I love my farmer" bumper stickers, but where's the love for local tech companies? Why? Because protective measures to keep intellectual property local are really the critical aspect in preparing for the next war our allied nations will fight. If not already apparent to the reader by the move towards non-physical battlefields such as Electromagnetic and Cyber, and the slew of emerging technologies discussed previously, the intellectual property surrounding technological advances is the key enabler for our armed forces. That information constitutes the criteria for win or lose in Phase 0 of war. There are many examples of planes, ships, trucks, unmanned vehicles, and missiles produced by foreign nations that look like street-corner knock-offs of allied systems, many of which are frightfully convincing [71,72]. To stop this, we must first secure that data, putting in place physical, personnel, and IT safety measures to prevent that data from being disseminated. But second, perhaps the largest overlooked vulnerability, is when the boss is an insider threat. Why hack a company and steal the information when a foreign entity can buy the company and bring the designs back to the native nation, while also making a profit and fostering more intellectual property? That's the icing on the cake, and this is happening at an alarming rate. As the US Pentagon reports, "China is investing in Silicon Valley start-ups with military applications at such a rapid rate that the United States government needs tougher controls to stem the transfer of some of America's most promising technologies[73]."

Problem four will be to press forward with our Defense Industry through the tidal wave of data. The explosion of information technology from Silicon Valley is producing more information and data, such that in 2010 Eric Schmidt, then CEO at Google, told a conference that as much data was being produced in 2 days now as was created from the beginning of humanity up to 2003[74]. To humanity's credit, this data wave is not causing an analysis paralysis, but indeed is forcing evolution. As this tsunami of information and data approaches, humans are now being forced to make more decisions, perform higher velocity learning, and update our stodgy linguistic norms with the supplementation of pictographs (emojis) to hasten and expand communication. But while humans adapt, our defense infrastructure must do the same. This acceleration will continue to antagonize the dawdling bureaucracy of the Defense Industry, forcing it to synthesize more information and data than it has bandwidth for, evolve the human-machine interface, evolve product pedagogy, and reduce fielding times for products before they become obsolete.

These are just 4 of 99 problems, but when you're fighting it, bureaucracy feels like 96 of them. When I'm done fighting it, I can only hope that I can be an old pensioner sitting in my rocking chair and reading a newspaper, or whatever incarnation of such a concept exists in 2050, and not have to worry about a foreign army blitzkrieging over the border like Red Dawn and raiding my New England town. Hopefully what the millennials have done to solve the 99 problems up to that point will prevent that. While the ex-servicemen portrayed in that Hollywood movie *Battleship* were overjoyed to return to service to fight aliens using an antique warship, and were successful in their mission, I feel that forcing my septuagenarian brain back into the business of designing new weapons for the future USS *Obama* would not be as glamorous and satisfying as the movies would make it out to be. So to prevent that from happening, let's discuss some ways to address these problems.

Chapter:6
Hack the Industry

"the only people for me are the mad ones…"
-Jack Kerouac,
American Author

No one envies the challenges millennials will face to maintain the national security of our allied nations. Thinking back to the millennials I watched leave my labs, I'd fight harder to keep them if I could do it over, because the challenge grows even harder every time another millennial walks out the door. Unlike some countries, we can't conscript citizens to work for the Defense Industry. You need to want to do it, it's voluntary, and if we can't convince our fellow citizens to join in, we don't stand a chance. So I'd plead with them that there's a lot of room for millennials to adapt their generational skillsets to the Defense Industry to make it a better place, and we can't overcome our national security challenges without them. In fact, one vocation millennials perform better than anyone is hacking. I'd tell them instead of leaving they should hack the Defense Industry, hack our 99 problems. What industry seems more vulnerable to a disruptive approach than the Defense Industry? A massive, monolithic, slow-paced, bureaucratic complex: ring a bell? Think of how upstarts to other mature industries like Amazon to retail, Uber to transportation, and AirBnB to hospitality changed how those industries function. There's a massive opportunity to do the same in the Defense Industry, and I know millennials could lead the way.

First hack, the Defense Industry can be a technically conservative place, so more creative safe spaces must be built for the freedom to develop new ideas and divergent thoughts[75]. In the US, the Defense Advanced Research Projects Agency (DARPA) (and its intelligence cousin IARPA) will always be safe spaces, but there's not enough like them. These are the "moonshot" ideation agencies for defense and intelligence that fund high-risk efforts primarily by commercial companies. DARPA has had some hand in the Internet, Siri, GPS, Roomba, and autonomous cars[76]. Recent projects include hacking the human brain, exoskeletons, hypersonic weapons, and the ability to climb up walls like Spiderman. Additional efforts by the US services to generate creative ideas include the Strategic Capabilities Office (SCO), Defense Innovation Unit (DIU), and Innovation Cells, which hasten the acquisition process from commercial companies while finding time to appear at events like SXSW[77]. These units also speak to hacking a more diversified creativity portfolio between service labs and industry. Think of good ideas like a craft beer, and the difference between how ales and lagers are made. Right now as the funding flows down from government to our military service labs, a great portion of R&D funding frequently stops there, top-fermenting creativity like an ale. There are advantages to that; those labs inherently have security clearances and they have closer relationships with the services and their needs. Much like an ale, this is easier to brew, the funding has shorter distance to move and can act faster, but has a higher risk of failure due to their lack of engineering and manufacturing experience. On the other hand, commercial companies at the bottom of the funding chain are the ones that truly design and manufacture the equipment, they are ones building developing new technologies, and they can contribute creativity of equal if not greater value. It requires a little more work to identify the right company, develop contracts, and help guide their creativity, but a well-lagered idea typically results in a more elegant, robust, and valuable defense product. This is one reason the US government used JASON, an independent group of scientists that advised the government, but the contract with the Department of Defense has recently ended[78]. That said, there are other encouraging signs, like the rise of defense technology specific incubators. In the state of Rhode Island, the privately funded Undersea Technology Innovation Center (UTIC) exists to help companies develop better marine products, primarily in support of undersea warfare[79]. Southern New England is a hotbed for undersea technology, from the world's premier submarine builder, to the premier unmanned underwater vehicle builders, to the navy's undersea warfare laboratory. Forming these sorts of defense incubators give start-ups a better chance of surviving by growing their business products in a community that recognizes the merit.

Hacks like these are a step in the right direction, but many of these new opportunities capture the interest of start-up companies or non-defense companies. It is time for existing defense companies to take a long, hard look at their internal investment. Perhaps the greatest potential of this hack is to convince executives and shareholders of existing Defense Industry companies to become more committed to sustainment and development. Performing more of this within the Defense Industry would help retain young employees. It's incredible to think of the amount of money within the Defense Industry, yet tech companies like Google and Apple spend up to five times more than the largest defense companies on internal R&D[80].

The second hack is to forget the war on the battlefield and focus on the war on the budget. When the defense budgets of allied nations are stressed by concurrently procuring multiple capital assets like new missile programs, joint strike fighters, ballistic nuclear submarines, aircraft carriers, and with pending big bills from the deferred maintenance of existing fleets and divisions that many allied nations conducted for the past decade, it's time to move risk away from complex new toys. A significant, and seemingly logical, hack to the Defense Industry would be to build simple, reliable products. As much as toy companies would want us to believe that kids are at their happiest interacting with an animatronic $500 puppy, we all know a cardboard box is all that's needed to spark overwhelming joy in a kid. Likewise, keep in mind in WWII when push came to shove, simple Liberty Ships were turned out like Model Ts in as short as a week or two. It's the hubris of person-kind that wants to continuously build more capability into the next widget. But "declining budgets in the Western world and growth in Asia and the Middle East give rise to an overwhelming trend in the Defense Industry: affordability. About 85 percent of executives believe that their customers will shift their focus from procuring systems with the highest possible performance to ones that are more affordable[81]." A component of this hack is for the military services to get woke on quality management and manufacturing, and the cost it takes to develop a high quality, manufacturable, and sustainable defense system. Designing a complex system is one cost, but the procurement and sustainment of a complex system is an immensely, significantly greater cost, a fact that many military acquisition personnel commonly forget. So before we hit the "Buy Now" button, we need to better understand the true time and cost to deliver that system.

However, it takes two to tango; there are times where our military services just want to "Buy Now" prototypes for testing without the considerations of a product's life cycle, and many companies lack the ability to bypass many of the quality management processes, their machinations entrenched in the requirements of larger defense contracts. So the fourth hack is the modularizing of defense contracts and the reciprocal modularity of company processes to allow for simplified acquisition. The structure of defense contracts flows down from the Defense Federal Acquisition Regulations (DFAR), something that looks related to our tax code. It gives no hint of being able to develop technology and services quickly, especially when it comes with calculated risk. Instead, defense companies and agencies take extremely conservative stances with contracts, including enough clauses and requirements to bore a lawyer. These clauses, while reducing risk, stifle the work being asked for, adding layers of work that provide no value in some scenarios. While many employees may complain about the "TPS Report Cover Sheet," they probably don't realize that it's probably a contractual requirement. Companies and the military services need a mechanism to agree on higher risk acquisition, quicker pace development and prototyping, less reporting, and best effort services, with lack of penalty or retribution (and equal blame!). With an element of "handshake" acquisition, and a little less legalese, companies could provide more of the services that the military acquisition community needs.

The fifth hack is to make sure you know the rules of the game better than your opponent: be clever. Military philosophy leans heavily on their "classics," including *History of the Peloponnesian War* by Thucydides, *The Prince* by Machiavelli, *The Art of War* by Sun-Tzu, and *On War* by Carl Von Clausewitz. We teach this knowledge to too few; I'm constantly amazed how few of my colleagues have never even heard of these, yet their jobs all trace to a lesson from one of these books. As Sun-Tzu stated, "what the ancients called a clever fighter is one who not only wins, but excels in winning with ease." But more so than how to win, these books give us insights on the prevention of war and protection of our freedoms, and thus explain our job requirements as well as any employee handbook.

A few other inevitable hacks will include wider implementation of modern workflow tools like Atlassian tools and Slack, increased use of model/virtual based product development to verify designs before they're built, and increased diversity[82]. To this last point, while our military services do an excellent job recruiting, the Defense Industry could certainly take a flyer on how to hack their recruiting for the civilian workforce and defense companies. A few more slick ads through targeted advertising channels could show teenagers and college kids the benefits of being part of the Defense Industry STEM workforce (assuming we can fix the process of getting them onboard like we discussed prior). Additionally, college business or engineering programs with a focus on the Defense Industry would also be of benefit to both sides before new graduates parachute into the Defense Industry oblivious to its rules. We have college programs that focus on the financial industry, the healthcare industry, but the Defense Industry is a unique segment unto itself, driven by unique requirements, defense acquisition laws, and unique product development processes. Defense Industry employees can perform even better if training starts earlier. Lastly, as these trained college graduates exit college, we must find ways to not only onboard them quicker, but find ways to facilitate a more fluid, freelancing work model so that the Defense Industry doesn't diverge any further from the growing gig economy model preferred by young generations.

Sometimes the challenge seems daunting; if the crazy despot with a finger on the nuclear missile button doesn't get you, maybe the frustration and despondency of battling the regulations and bureaucracy of the Defense Industry will. But with the right hacks, the Defense Industry can be less daunting and an awesome place to work for almost anyone. Almost? I can't sell it to everyone, hacks or no hacks, working in the Defense Industry will always require a certain wherewithal, grit, and madness. Why? Because life in the Defense Industry is a grind, it's a struggle against threats to our national security, it's a struggle against bureaucracy, it's a struggle to hack the bureaucracy, and it's a struggle to stand up to people telling you that you can't design a widget that improves our national security. It doesn't matter if it's social media, sports, or STEM fields, there will always be haters. So it takes a little madness to rise above the noise and to stay determined and focused. But if you can commit to it, making a career in the Defense Industry is one of the best choices you can make in life, and our allied nations will be better for it.

Chapter:7
Netflix and Defend

"All the world's a stage and most of us are desperately unrehearsed"

-Sean O'Casey,
Irish Dramatist

As movies like *Captain America* and *Independence Day* demonstrate, Hollywood has mined (no pun intended) the Defense Industry for entertainment for decades. Similarly, the relationship between the defense complex and the beverage industry has been strong since the beginning of military history. Viking raid victory parties would have been a joyless affair without a good tipple of mead. So while the first chapters were a buzz kill, the flipside of the Defense Industry is amazing. To prove it, in lieu of a humdrum list of examples, we can Netflix it alongside some themed beverages.

The playlist kicks off with lighter fare. First, a comedy that had a lasting effect on inner-geeks and propelled many towards a job in technology was the 80s comedy "Real Genius." How can our Defense Industry tap into uber-IQ, young Einstein geniuses and bypass their moral calculus of building revolutionary lethal capabilities for defense end-item use? Lie to them! Some similarities can be drawn between this 80s comedy classic and *Ender's Game* (novel or movie), where young geniuses are used to unknowingly commit alien genocide, but this comedic variant lends itself better to mornings. As an alternative, or up second, the *Pentagon Wars* is a must-watch for anyone that works in the Defense Industry; in fact we even watched this in class at the US Naval War College. A cynical, *Office Space*-like takedown of the defense complex, particularly regarding how self-interest and politics are unfortunately never divorced from the process of fielding equipment for the people that are dodging bullets to protect our freedoms.

Assuming an early start, the drink pairing goes one of two ways: recovery, or straight hair of the dog. First, for recovery and the gold standard, a cup of coffee. Lore is that Americans started drinking more coffee following the decrease in tea supplies after the Boston Tea Party, and it's been fueling our defense complex ever since. But to give credit back to America's old foe, when I was working with the Royal Navy and UK Defense Science and Technology

Laboratory, I relied on Lucozade in my hotel mini-bars, a unique British energy drink with just massive amounts of sugar and caffeine, and trace levels of alcohol. Everything a body needs in the morning. However, if coffee or Lucozade can't make us feel better about last night's drunk texting and tweeting, let's travel to the GIUK (Greenland-Iceland-United Kingdom) gap. The GIUK gap is the famous ocean path between Soviet naval bases and the Atlantic Ocean that brought many a NATO serviceman and woman through Iceland, and probably an exposure to Iceland's top spirit, Brennivin. Roughly translating via various Scandinavian languages as 'burning wine' and also referred to as 'black death,' a sip of Brennivin in the morning will get you right.

After the laughs, the mid-day tone gets more adulting, with movies such as *Imitation Game* and *Hidden Figures*, two movies in the theme of government funded science that highlight the magnanimity of pursuing engineering achievements in aerospace and military operations in the face of bias. They help depict the universal truth that genius is a human attribute, blind to race, sex, or sexual preference—military STEM is not an old, straight, white man's game. After these earnest moments and a long day of binge watching, it's time to relax. In fact, the origin of the phrase "happy hour" comes from the US Navy. The crew of the USS *Arkansas* is credited with the phrase for holding "happy hours" on their ship back in 1913, regularly scheduled entertainment events, a custom that spread through the US Navy by the end of World War I[83]. Later in the century, another world war dragged millions of servicemen and women across the Pacific, resulting in the post-war tiki culture fad, which presented us with one of the best happy hour drinks, the Mai Tai.

After happy hour, bring it home with a few action-drama blockbusters, perhaps a combination of *War Games* and the *Matrix*. *War Games* is the grand-daddy of cyber films, a prescient 1983 movie about hacking defense networks and its potential role in warfare. A generation later came the *Matrix* trilogy, less to do with conventional warfare, but perhaps more than other movies it brought to the forefront of societal consciousness that cyber, just like space, air, ground, and sea, is in fact a domain unto itself that has to be defended. Then, as darkness falls, imagine the darkness of the deeps with *Hunt for Red October*. The mountaintop of human engineering ingenuity is best characterized by a few machines, including spacecraft, aircraft carriers, and nuclear submarines. Combining the drama of Cold War political chess with the intriguing notion of an invisible submarine drawn from science fiction (or is it from reality? You'll need security clearance to find out), a mutiny, and the action and tension of a large screen, this war feature deservingly remains on rotation on cable TV to this day.

Playlist exhausted, we leave Netflix behind and curl up with Kindle. A permanent spot on my bookshelves is always reserved for *Blind Man's Bluff, The Right Stuff*, and any book on Area 51, the sort of stories about highly classified or highly complex feats of engineering that excites the Legobuilder kid in me. Of course the moral complicity of participating in defense programs must always be considered, and John Hersey's book *Hiroshima* on the stories of six survivors of the atomic bomb was one of the best journalistic efforts of the 20th century, and puts in the context the responsibility that freedom engineers must consider. In 1945, Oppenheimer famously quoted the Hindu text Bhagavid- Gita, "Now I am become death, the destroyer of worlds" following the detonation of a nuclear weapon he helped devise. The moral struggle of pulling a distant trigger, or creating the opportunity to put the trigger in the hands of a stranger, will always be an anxious decision that every Freedom Engineer must come to peace with. Ultimately, the Freedom Engineer must concede their job exists for a singular purpose, as said best by General MacArthur, "Our mission remains fixed, determined, inviolate. It is to win our wars. Everything else in your professional career is but a corollary to this vital dedication."

Chapter:8
Freedom Engineers of Instagram

*"The question isn't who is going to let me;
it's who is going to stop me"*
-Ayn Rand (paraphrased),
Russian-American Writer

We all have influences in our lives that lead us towards our goals. In today's world, this is synonymous with Instagram, filtered depictions of the lives or jobs we want. The Defense Industry doesn't lend itself well to Instagram, security being one significant reason. It's hard to advertise the merits of an engineer or scientist that succeeded in developing groundbreaking classified equipment without creating a security incident and revealing our secrets. So part of finding an inspiring Freedom Engineer requires one to be one first. It's a leap of faith required in joining the Defense Industry, but I'll guarantee one that pays off.

I encountered dozens of influencers in my defense career. My favorite was a successful navy officer that I collaborated with, who at one time gave me the opportunity to stand next to him on the bridge of his ship as we pulled into Pearl Harbor. As we stood there, he explained to me the sequence of events and frantic movement of ships to avoid the Japanese torpedoes during the attack on Pearl Harbor, all while his nuclear powered ship slid through the narrow harbor channel into its berth. Amidst what seemed to me was a high stress event of parking his multi-billion dollar responsibility, he made me feel like a VIP. That experience alone kept me in the navy's employment for years after, even as I had opportunities in other industries.

While we can't always confirm or deny the merits of defense influencers, we can look to ancient history, recent history, and even fictional figures to help us paint a picture of what it's like. Archimedes for instance, one of the greatest mathematicians of ancient Greece and all of history, is also credited for inventing catapults to ward off the Romans from his home city of Syracuse. Leonardo Da Vinci, while not painting Mona Lisa, practiced as a military engineer, proposing advanced defense technology ideas from scuba gear for "frogmen" to disable enemy ships from beneath, to multi-barrel cannons.

In more recent history Alan Turing, portrayed in a recommended movie from last chapter, was a brilliant British scientist who was responsible for cracking German codes during World Word II, inventing the computer, and also the

"Turing Test," a standard to determine if a machine was artificially intelligent. Stateside, Kelly Johnson was the famous founder and leader of the Lockheed Skunk Works, the famous classified lab that designed aircraft like the U-2 and SR-71, and probably a slew of amazing aircraft being tested at Area 51 that we won't know about for decades. Down at sea level, Admiral Rickover of the US Navy was best described in his *New York Times* obituary as having "attacked Naval bureaucracy, ignored red tape, lacerated those he considered stupid, bullied subordinates and assailed the country's educational system. And he achieved, in the production of the nuclear-powered submarine in the early 1950's, what a former Secretary of the Navy, Dan Kimball, called 'the most important piece of development work in the history of the Navy[84].' " The nuclear ships, first submarines and then aircraft carriers, that were created under his watch during his 63 years of active service in the navy (at the time a navy service record) created a dominant and globally present US Navy. From an engineering perspective alone, nuclear naval ships are perhaps the most technically complex products humanity has ever produced.

The list of fictional figures is long and entertaining. Two examples repeatedly on the big screen due to their movie franchise success include Tony Stark from *Iron Man* and Q from *James Bond*. Q is more emblematic of the Freedom Engineer or scientist, operating behind the curtain to provide enabling technologies for 007 to accomplish his objectives, including x-ray glasses, a car cloaking device, submarine cars, and jetpacks. Tony Stark built the next-next-generation Amazon Alexa in JARVIS, and of course the Iron Man outfit and every gadget he integrated into the suit.

Specific examples of Freedom Engineers aside, every day millions of current freedom influencers get up and go to work and grind every day in the Defense Industry. But we haven't discussed *what* organizations actually make up the Defense Industry and *where* the Defense Industry actually is! Is it specific companies or specific parts of our defense organization? What if you are interested in being part of the Defense Industry, but are a biologist and not a software engineer? What if you live in Indiana and not Washington DC? The great thing is that there is an organization with the Defense Industry that fits anyone's needs.

Defense Industry organizations range from very academic institutions to very industrial, from very classified (some technically don't exist) to very unclassified, and some are commercial private companies, while some are federal organizations, and some are non-profit. Some of the well-known companies include BAE Systems, Lockheed Martin, the Boeing Company, Thales, Raytheon, General Dynamics, Northrop Grumman, Kongsberg, United Technologies, and L3 Technologies. There are literally thousands of smaller defense companies, many in every state or territory. Alternatively, some of the well-known federal defense workplaces include the NATO Science and Technology Organization, various US DoD science and technology organizations such as the Naval Research Lab; Office of Naval

Research; Air Force Research Laboratory; US Army Research, Development, and Engineering Command; Naval Sea Systems Command; Naval Undersea Warfare Center; Naval Information Warfare Center; Defense Advanced Research Projects Agency; Intelligence Advanced Research Projects Agency; and the National Security Agency, or our friends in the US Coast Guard Research Center. Amongst our allies, you might work with folks from the Australia Defence Science and Technology Group and the UK Defence Science and Technology Laboratory. At yet another organizational level, there are US DoD university affiliated research centers (UARCs), some of the most famous being Johns Hopkins Applied Physics Laboratory, Penn State Applied Research Laboratory, University of Washington Applied Physics Laboratory, University of Texas Applied Research Laboratory. And then another type of organization, US federally funded research and development centers (FFRDCs), of which there are many, but some of the more well-known include MIT Lincoln Laboratories, Jet Propulsion Laboratory, Lawrence Livermore National Laboratory, and the Software Engineering Institute.

With all of these organizations, any engineer or scientist has an opportunity to follow their interests, whether it's specific areas of research (from cyber to biology), specific geographical areas (from earth to space), specific salary grades (competitive to 1%), or a specific type of work environment (from sandals to brogues). The Defense Industry provides a Goldilocks option for all. This versatility makes the Defense

Industry one of the most unique and rewarding sectors for a career, and thus it's not surprising that so many engineers and scientists become great leaders and influencers, because they can find rewarding and satisfying career paths, and opportunities to be successful.

Chapter:9
The Future - Generation Z

"The future has several names. For the weak, it is impossible; for the fainthearted, it is unknown; but for the valiant, it is ideal"

-Victor Hugo,
French Novelist

Part of being a Freedom Engineer means staying two steps ahead of our adversaries, which includes not only training your replacement, but also considering your replacement's replacement. Therefore an introduction is needed for the up-and-coming Generation Z ("Gen Z" for short). They are millennials' younger siblings or children, born in the late 90s and early 2000s, and have a unique view of the world based on a childhood in the economic recession and being more digitally embedded than any previous generation, renowned to have high expectations, and have even more entrepreneurial spirit than millennials[85].

But for the millennial frustrated with this whipper-snapper Gen Z, I suggest just to wait a moment, and... squirrel! Gen Z is widely regarded as the least focused and easiest distracted, due to the deluge of digital distractions they have been raised with. This creates a significant concern with Defense Industry work based on the complexity and tedium of bringing a defense design to fruition, as we've previously discussed, with programs stretching into decades. This begs the question of whether we can we rely on Gen Z to make sound decisions without the level of steadfastness of previous generations. National security decisions are often quite complex and require focus; in fact at the US Naval War College there is an entire course of study in National Security Decision Making. The NSDM "course integrates concepts drawn from a number of disciplines including international relations, area studies, foreign policy analysis, leadership studies, and cognate fields[86]." No one is able to perform that level of decision making without the time and experience it takes to accumulate expertise in these fields, which may be challenging with Gen Z's wandering focus.

Additionally, while the participation trophy cliché was born in the millennial generation, it was institutionalized in the Gen Z generation, and this emphasis on attempt over achievement will have troubling implications for leadership and worker bees in the Gen Z Defense Industry. The Defense Industry requires huge numbers of participants to fall in line and thanklessly execute large integrated engineering and management plans successfully. This quantity of non-affirmative, result-driven work might bristle the participation trophy-leaden Gen Z worker. Furthermore, they'll need to not only forego the lack of affirmation; they'll also need to forego the need to establish leadership and work to be a better follower. Sociological theory is suggesting that society indeed needs more followers than leaders, "a discipline in organizational psychology, called 'followership,' is gaining in popularity, as less and less of the population identify with being a follower and more identify with being a leader due to leadership and accomplishment glorification. Robert Kelley, a professor of management and organizational behavior, defined the term in a 1988 *Harvard Business Review* article, in which he listed the qualities of a good follower, including being committed to 'a purpose, principle or person outside themselves' and being 'courageous, honest and credible.' It's an idea that the military has long taught[87]."

On the positive side, perhaps one of the most appealing aspects of Gen Z is they tend to be frugal because of their childhood in recession, and no one needs more fiscal caution than the Defense Industry, especially as global defense spending growth is near neutral. Additionally, "(Gen Z) are the first true digital natives[88]." In doing so, there's great confidence they will be the first true cyber-woke-adults, the first generation that takes the digital world with well-measured, mature, and secure responsibility. They have pulled back from the almost hippie free-love embrace of digital data sharing on the internet, like late Gen Xers and millennials conducted over the course of their lives. Gen Zers have in fact gravitated to more secure and anonymous tools like Whisper and Snapchat, and are more sensitive to privacy concerns generated from Wikileaks and the Snowden incident. This high level of digital responsibility translates directly to the battlefield, keeping our allied nations safer in the future through better cybersecurity.

Ultimately I would propose this transition of power from Gen Millenial to Gen Z will be even more critical than the current transition from Gen X to Gen Millenial. Our current handover to the millenials is occurring during relatively stable times. Regardless of what theory of international relations you subscribe to, there is an indisputable rise of new powers and forces as described in Chapter 3 that will certainly erode the power and influence of our allied nations in the future, creating levels of tensions unseen since the Cold War, just as Gen Zers are filling roles in the Defense Industry.

Alternatively, the possibility exists that Gen Z will step into these roles and dismiss these tensions as the ravings of old generations, and find methods to de-escalate these situations and achieve peace dividends once again. Due to being digital natives, the ongoing breakdown of identification (gender, sexuality, etc.), and their job-hopping parents, no generation will have been as globalized and inclusive as Gen Z by the time they're in adulthood. Ultimately, perhaps the generation will have the framework to view the world in a way where hawkish, divisive tensions may be irrelevant. Hope isn't a strategy, but as much we love being Freedom Engineers and we love the Defense Industry, our greatest hope should always be that we're not needed, and perhaps Gen Z will be the first generation to make that happen.

Chapter:10
Here are the Keys

"If you had access to a car like this, would you take it back right away? Neither would I."

-Ferris Bueller,
Movie Character

In 2017 a guy in Australia compiled a comprehensive list of all the things reported in the media that millennials have supposedly "killed." Vacations, running, committed relationships, and even entire product types, including cereal, movies, napkins, and golf[89]. But what about the Defense Industry? Will it become the Defense Millendustry, or just another thing killed by millennials? Though I've witnessed their frustration with the Defense Industry, I don't believe millennials will kill the Defense Industry, I believe they'll take ownership of the Defense Industry and build a Defense Millendustry that's better than the one we handed them.

However, when it is time for millennials to take that janitor-sized Defense Industry key chain from their Gen X peers, it'll come with our expectations that they'll be highly trained, skilled, and proficient at their jobs. It is the responsibility of those that lead the Defense Industry to navigate it safely, as it not only protects our national security and our freedoms, but also fuels our economy like no other industry. Consider that in the US alone, the DoD is approximately 3% of the national budget, over $600 billion, which is the largest single component of the budget and approximately 50% of the discretionary portion of the budget[90]. For context, that'd buy about 40 million bomber bottles of craft beer, which could stack up 5.28 billion feet, enough to go to the moon and back, and back to the moon. In some states the defense sector comprises up to 13% of the state output, and the defense sector consists of 10-20% of the national research and development funding[91]. The US DoD federal workforce, not including industry, is the largest employer in the world[92]. When you include the commercial Defense Industry, the number doubles to approximately 6 million jobs in the US alone[93]. When you include allied nations, federal and private workforce, it accounts for over 7.5 million jobs worldwide.

More importantly, a portion of these 7.5 million jobs across our allied nations are involved in nuclear weapon research and control a stockpile of over 2400 active nuclear weapons, and a stockpile of thousands more[94]. Our adversaries control thousands as well, and more countries (and non-state groups) want to join the list of nuclear weapons

holders. The nuclear sector is a job that transcends generations, a job that my generation didn't start but inherited and
safeguarded, and it'll be a job that millennials will inherit,
and it comes with the responsibility to keep our nuclear
technology and weapons secure, the responsibility to peri-
odically refresh the technology while maintaining its security, and ultimately the responsibility to use it if necessary.
The access to these missile keys is the most critical component of protecting democracy and humanity, and it's a key
on the keychain that will weigh disproportionately heavy in
the skinny jean pockets of our new bosses.

Moreover, millennials should remember that their
fellow citizens and taxpayers are providing their hard
earned money for their job in the Defense Industry, and so
working in the Defense Industry means being accountable
to everyone you know, all your friends, family, and neigh-
bors. It also means being accountable to everyone you'll
meet, because the Defense Industry provides a great opportunity to travel. In fact, if you combine normal work travel
with some symposiums and conferences, defense work can
be as jet-setting as you'd like. No other career path I would
have chosen would have allowed me the opportunity to boat
from a nuclear submarine to a beach bar on a summer's
night, or the opportunity to stumble upon rare vinyl of my
favorite band in the Scottish Highlands, or happen to sit behind one of my other favorite bands on a plane to Australia,
or to get a week's worth of dinners from late night food stalls
in South Korea, or go surfing on my lunch break in San

Diego, or dive on WWII wrecks on a Pacific atoll. In each of these places, in each of these experiences, I had to speak to strangers about where I worked and why I was there, and I did it with a sense of pride.

From a different perspective, what alternative do millennials have? You might think the millennial generation would be start-up drunk, as popularized in pop culture. But keep in mind that statistics on tech startup failures are soul crushing, with over 90% failure[95]. Startups typically result in low pay, short-term jobs, with little to show after it, as the company is rarely successful and the high-risk payout possibility falls through. In fact, large companies can on average pay more than Silicon Valley, without the high risk of job loss[96]. Additionally, some argue that over the long run, when you look at long-term benefits, you earn even 78% more in benefits in federal Defense Industry jobs[97]. Certainly the appeal of start-ups is real, and their zany environments, but as we've covered, the Defense Industry is so highly segmented that anyone can find an environment that suits them, and that's before we hack the industry to make it an even better place to work. There is also something about being part of the largest fraternal and sororal order of STEM employees working as scientists, mathematicians, and doctors to support our freedoms and will provide benefits and camaraderie throughout your life.

So as the silver tsunami swells and we approach a new era of contentious international relations, the Defense Industry and its ranks of new Freedom Engineers will play a critical role in our future, regardless if it's called the Defense Industry, Millendustry, or Zindustry. While I may have witnessed some initial hesitation, I have no doubt millennials will fill these ranks. I have no doubt our new millennial bosses will have the intelligence, creativity, drive, and madness to lead the Defense Industry and overcome its problems (along the way their robots may even learn how to play foosball and defeat us humans, so maybe we shouldn't have foosball tables at our lab anyhow). I have no doubt Gen Z will take the baton from them in the future and make the Defense Industry even better. And we can try to perform expensive decoding of generations, but there is nothing to decode in this: Freedom Engineers protect our democracy and freedoms, serving your country transcends generational divides. Godspeed Defense Millendustry.

Notes

[1] Gellman, L. (2016, 5 16). Helping Bosses Decode Millennials— for $20,000 an Hour. Retrieved 4 14, 2017, from The Wall Street Journal: https://www.wsj.com/articles/helping-bosses-decode-millennialsfor-20-000-an-hour-1463505666

[2] Giesener, J. (n.d.). *Why is Every Candidate Expecting Mobile Access to Jobs?* Retrieved 4 15, 2017, from Source Mob: http://sourcemob.com/why-is-every-candidate-expecting-mobile-access-to-jobs/

[3] Poswolsky, A. S. *"What Millennial Employees Really Want,"* Fast Company, https://www.fastcompany.com/3046989/what-millennial-employees-really-want

[4] Greenwell, J. (2016, 5 20). *how to Attract Millennials to the Public Sector.* Retrieved 4 14, 2017, from Government Executive: http://www.govexec.com/excellence/promising-practices/2016/05/how-attract-millennials-public-sector/128276/

[5] Mussell, R. M. (1996). *Changes in Federal Civilian.* Congressional Budget Office. Washington DC: Congressional Budget Office.

[6] Zillman, C. (2013, 11 12). *America's defense industry is going gray.* Retrieved 4 15, 2017, from Fortune: http://fortune.com/2013/11/12/americas-defense-industry-is-going-gray/

[7] Wikipedia. (2017, 12 10). *Military budget of the United States.* Retrieved 12 11, 2017, from Wikipedia: https://en.wikipedia.org/wiki/Military_budget_of_the_United_States

[8] Institute for College Access and Success (TICAS). (2016, 10 1). *Student Debt and The Class of 2015, 11th Annual Report.*

Retrieved 4 14, 2017, from Institute for College Access and Success (TICAS): http://ticas.org/sites/default/files/pub_files/classof2015.pdf

[9] Yoder, E. (2016, 10 31). *Federal Employees behind in pay by 34% on average, salary council says.* Retrieved 4 116, 2017, from Washington Post: https://www.washingtonpost.com/news/powerpost/wp/2016/10/31/federal-employees-behind-in-pay-by-34-percent-on-average-salary-council-says/?tid=a_inl&utm_term=.d7a514bc3167

[10] Poswolsky, A. S. (2015, 6 4). *What Millennial Employees Really Want.* Retrieved 4 15, 2017, from Fast Company: https://www.fastcompany.com/3046989/what-millennial-employees-really-want

[11] Fromm, J. (2015, 11 6). *Millennials in The Workplace; They Don't Need Trophies But They Want Repinfor.* Retrieved 4 14, 2017, from Forbes: https://www.forbes.com/sites/jefffromm/2015/11/06/millennials-in-the-workplace-they-dont-need-trophies-but-they-want-reinforcement/#74c0ae4953f6

[12] Adkins, A. (2016, 5 12). *Gallup.* Retrieved 4 14, 2017, from Millennials: The Job-Hopping Generation: http://www.gallup.com/businessjournal/191459/millennials-job-hopping-generation.aspx

[13] Giesener, J. (n.d.). *Why is Every Candidate Expecting Mobile Access to Jobs?* Retrieved 4 15, 2017, from Source Mob: http://sourcemob.com/why-is-every-candidate-expecting-mobile-access-to-jobs/

[14] Ganpathy, S. (2016, 9 1). *10 Millennial Personality Traits That HR Managers Can't Ignore.* Retrieved 4 14, 2017, from

MindTickle.com: https://www.mindtickle.com/blog/10-millennial-personality-traits-hr-managers-cant-ignore/

[15] Florida, R. (2015, 4 20). *America's Leading Creative Class Cities in 2015.* Retrieved 5 1 2016, from City Lab: http://www.city-lab.com/work/2015/04/americas-leading-creative-class-cities-in-2015/390852/

[16] Military.com. (2017). *Top 10 Cities for Defense Jobs.* Retrieved 4 16, 2017, from Military.com: https://www.military.com/veteran-jobs/search/aerospace-defense-jobs/top-defense-job-cities.html

[17] Suciu, P. (2016, 5 23). *Best Cities for Defense and Intelligence Jobs.* Retrieved 4 16, 2017, from ClearaanceJobs.com: https://news.clearancejobs.com/2016/05/23/best-cities-defense-intelligence-jobs/

[18]UniqueHR.com (2016, 11 4). *Stop Fighting the Future: Marking Millennials Part of the Team.* Retrieved 7 2, 2017, from Unique HR: https://www.uniquehr.com/stop-fighting-the-future-making-millennials-part-of-the-team/

[19] Whitlock, C. and Woodward, B (2016, 12 5). *Pentagon buries evidence of $125 billion in bureaucratic waste.* Retrieved 6 10, 2017, from The Washington Post: https://www.washingtonpost.com/investigations/pentagon-buries-evidence-of-125-billion-in-bureaucratic-waste/2016/12/05/e0668c76-9af6-11e6-a0ed ab0774c1eaa5_story.html?utm_term=.119b3a73c77b

[20] Fisher, A. (2016, 6 15). *Millennial Employees Could Be Your Company's Biggest Cybersecurity Risk.* Retrieved 9 17 2017, from Fortune: http://fortune.com/2016/06/15/millennial-employees-cybersecurity-risk/

[21] Moore, K. (2014 ,10 2). *Millennials Work for Purpose, Not Paycheck.* Retrieved 4 14, 2016, from Forbes:

https://www.forbes.com/sites/karlmoore/2014/10/02/millenni-als-work-for-purpose-not-paycheck/#62df5b46a51f

[22]Betz, B. (2017, 12 23). *'There's a war coming,' Marine Corps general warns US troops*. Retrieved 12 26 2017 from Fox News: http://www.foxnews.com/us/2017/12/23/theres-war-coming-marine-corps-general-warns-us-troops.html

[23]Chan, S. (2018, 1 24). *Doomsday Clock is Set at 2 Minutes to Midnight, Closest Since 1950s*. Retrieved 2 2 2018, from New York Times: https://www.ny-times.com/2018/01/25/world/americas/doomsday-clock-nu-clear-scientists.html

[24] Office of Secretary of Defense (2018, 5 16). *Annual Report to Congress: Military and Security Developments Involving the Peo-ple's Republic of China 2018*. Retrieved 5 20 2018, from Depart-ment of Defense: https://media.de-fense.gov/2018/Aug/16/2001955282/-1/-1/1/2018-CHINA-MILI-TARY-POWER-REPORT.PDF

[25] Saunders, P.C. and Chen, J. (2016, 10 1). *Is the Chinese Army the Real Winner in PLA Reforms?*. Retrieved 04 15 2017, from National Defense University Press: http://ndupress.ndu.edu/Me-dia/News/Article/969659/is-the-chinese-army-the-real-winner-in-pla-reforms/

[26] Gill, M.S. "Merchant of Dreams." Islands Magazine, July-Au-gust, 1993.

[27] Mullen, J. (2017, 1 9). *China posts weakest annual economic growth in 26 years*. Retrieved 1 25 2017, from CNN: http://money.cnn.com/2017/01/19/news/economy/china-fourth-quarter-gdp-economic-growth/

[28] Carpenter, T.G. (2016, 9 6). *Could China's Economic Troubles Spark a War?* Retrieved 8 3 2018, from The National Interest:

https://nationalinterest.org/feature/could-chinas-economic-troubles-spark-war-13784

[29] Einstein, J. (2017 1 17). *Economic Interdependence and Conflict – The Case of the US and China*. Retrieved 1 25 2017, from E-International Relations: http://www.e-ir.info/2017/01/17/economic-interdependence-and-conflict-the-case-of-the-us-and-china/

[30] The Editorial Board (2019, 2 10). *For Decades, the United States and Russia Stepped Back from the Brink. Until Now*. Retrieved 2 10, 2019, from The New York Times: https://www.nytimes.com/2019/02/10/opinion/trump-putin-inf-treaty.html

[31] *Nuclear Weapons: Who has What at a Glance*. Retrieved 8 1 2018, from Atrms Control Association: https://www.armscontrol.org/factsheets/Nuclearweaponswhohaswhat

[32]Oliphant, R. (2015, 5 6). How *Vladimir Putin's military firepower compares to the West*. Retrieved 4 12 2017, from the Telegraph: http://www.telegraph.co.uk/news/worldnews/vladimir-putin/11586021/How-Putins-military-firepower-compares-to-the-West.html

[33]The Office of Naval Intelligence (2015, 12). *The Russian Navy, a Historic Transition*. Retrieved 04 14 2017, from USNI News: https://news.usni.org/2015/12/18/document-office-of-naval-intelligence-report-on-russian-navy

[34]Scuitto, J. (2016, 4 19). Top Navy Official: *Russian sub activity expands to Cold War level*. Retrieved 4 16 2017, from CNN: http://www.cnn.com/2016/04/15/politics/mark-ferguson-naval-forces-europe-russian-submarines/

[35]Gady, F. (2015, 3 3). *Russia Tops China as Principal Cyber Threat to US*. Retrieved 8 4 2017, from The Diplomat: http://thediplomat.com/2015/03/russia-tops-china-as-principal-cyber-threat-to-us/

[36] Sharkov, D. (2017, 1 4). *NATO: Russian aircraft intercepted 100 times above Baltic in 2016*. Retrieved 1 7 2017, from Newsweek: http://www.newsweek.com/nato-intercepted-110-russian-aircraft-around-baltic-2016-538444

[37] Campbell, S. (2017, 5 27). *Pair of RAF Typhoon fighter jets are scrambled to intercept two Russian planes entering UK airspace as Putin ratchets up tension with the West*. Retrieved 5 27 2017, from DailyMail.com, http://www.dailymail.co.uk/news/article-4547530/RAF-Typhoon-fighter-jets-scrambled-Lossiemouth.html

[38] Gady, F. (2017, 1 26). Japan Scrambles Fighter Jets to Intercept 3 Russian Strategic Bombers. Retrieved 8 4 2017, from The Diplomat: http://thediplomat.com/2017/01/japan-scrambles-fighter-jets-to-intercept-3-russian-strategic-bombers/

[39] Durden, T. (2016, 10 19). Russia is Deploying The Largest Naval Force Since The Cold War For Syria: NATO Diplomat. Retrieved 7 13 2017, from Zero Hedge: http://www.zerohedge.com/news/2016-10-19/russia-deploying-largest-naval-force-cold-war-syria-nato-diplomat

[40] Gallagher, S. (2017, 2 15). Up close and personal: Russian spy ship skims edge of US waters near sub base. Retrieved 2 19 2017, from ARS Technica: https://arstechnica.com/tech-policy/2017/02/up-close-and-personal-russian-spy-ship-skims-edge-of-us-waters-near-sub-base/

[41] Axe, D. (2016, 6 27). *Iran's Trying to Rebuild Its Air Force*. Retrieved 6 18 2017, from The Daily Beast: http://www.thedailybeast.com/articles/2016/06/27/iran-s-trying-to-rebuild-its-air-force.html

[42] McCafferty, G. (2016, 10 10). *Anniversary parade provides rare glimpse into North Korea's military might*. Retrieved 5 26 2017, from CNN: http://www.cnn.com/2015/10/09/asia/north-korea-military-might/

[43] Amadeo, K. (2019, 3 15). *War on Terror Facts, Costs and Timeline*. Retrieved 3 20 2019, from the balance: https://www.the-balance.com/war-on-terror-facts-costs-timeline-3306300

[44]Institute for Economics and Peace (2015). *Global Peace Index 2015*. Retrieved 10 14 2018, from Relief Web: https://reliefweb.int/report/world/global-peace-index-2015

[45] Carothers, T. and Youngs, R. (2015, 10 8). *The Complexities of Global Protests*. Retrieved 6 10 2017, from Carnegie Endowment for International Peace: http://carnegieendowment.org/2015/10/08/complexities-of-global-protests-pub-61537

[46] Cillizza, C. (2017, 4 8). *This is the most amazing chart on Congress you'll see today*. Retrieved 5 9 2018, from CNN: http://www.cnn.com/2017/04/07/politics/house-swing-seats-congress/

[47] International Monetary Fund (2017, 1). *World Economic Outlook Update: A Shifting Global Economic Landscape*. Retrieved 9 6 2017, from International Monetary Fund: https://www.imf.org/external/pubs/ft/weo/2017/update/01/

[48] Miller, C. (2017, 3 28). *Evidence That Robots Are Winning the Race for American Jobs*. Retrieved 10 10 2017, from The New York Times: https://mobile.nytimes.com/2017/03/28/upshot/evidence-that-robots-are-winning-the-race-for-american-jobs.html

[49] Research and Markerts (2017, 3 13). *$226.2 Billion Robotics Market Forecasts, 2021: Consumer, Enterprise, Industrial Healthcare & Military Robots, Unmanned Aerial Vehicles and Autonomous Vehicles – Research and Markets*. Retrieved 7 16 2017, from Research and Markets, http://www.prnewswire.com/news-releases/2262-billion-robotics-market-forecasts-2021-consumer-enterprise-industrial-healthcare--military-

robots-unmanned-aerial-vehicles-and-autonomous-vehicles---research-and-markets-300422451.html

[50] Mehta, A. (2017, 3 30). *DoD Weapons designer: Swarming teams of drones will dominate future wars*. Retrieved 8 13 2017, from Defense News: http://www.defensenews.com/articles/dod-weapons-designer-swarming-teams-of-drones-will-dominate-future-wars

[51] Pomerleau, M. (2017, 3 30). *Cyber Command now looking to equip its cyber warriors*. Retrieved 5 20 2017, from Fifth Domain: http://fifthdomain.com/2017/03/30/cyber-command-now-looking-to-equip-its-cyber-warriors/

[52] Faram, M. (2016, 6 19). Navy forges ahead with plan to hire civilians for chief, captain. Retrieved 3 20 2017, from Navy Times: https://www.navytimes.com/story/military/2016/06/19/navy-forges-ahead-plan-hire-civilians-chief-captain/85877884/

[53] Natter, A. and Chediak, M. (2017, 1 6). *U.S. Grid in 'Imminent Danger' From Cyber-Attack, Study Says*. Retrieved 9 3 2017, from Bloomberg: https://www.bloomberg.com/news/articles/2017-01-06/grid-in-imminent-danger-from-cyber-threats-energy-report-says

[54] Wilson, J.R. (2016, 8 1). *Today's battle for the electromagnetic spectrum*. Retrieved 10 5 2017, from Military and Aerospace: http://www.militaryaerospace.com/articles/print/volume-27/issue-8/special-report/today-s-battle-for-the-electromagnetic-spectrum.html

[55] Freedburg, S.J. Jr. (2014, 10 10). Navy Forges New EW Strategy: Electromagnetic Maneuver Warfare. Retrieved 4 6 2019, from Breaking Defense: http://breakingdefense.com/2014/10/navy-forges-new-ew-strategy-electromagnetic-maneuver-warfare

[56] *Militarisation of space*. Retrieved 8 20 2017, from Wikipedia: https://en.wikipedia.org/wiki/Militarisation_of_space#Space_treaties

[57] Daniels, J. (2017, 3 29). *Space arms race as Russia, China emerge as 'rapidly growing threats' to US*. Retrieved 7 25 2017, from CNBC: http://www.cnbc.com/2017/03/29/space-arms-race-as-russia-china-emerge-as-rapidly-growing-threats-to-us.html

[58] Army Technology, 2013, 4 14. *Creating Supermen: battlefield performance enhancing drugs*. Retrieved 3 13 2018, from Army Technology: http://www.army-technology.com/features/featurecreating-supermen-battlefield-performance-enhancing-drugs/

[59] Larter, D. (2017, 5 16). *'Performance enhancing drugs' considered for Special Operations soldiers*. Retrieved 2 10 2018, from Defense News: http://www.defensenews.com/articles/special-operations-command-wants-to-develop-super-soldiers

[60] Diamond, J. and Starr, B. (2016, 4 11). *U.S. military is on its way to getting its Iron Man*. Retrieved 6 23 207, from CNN: http://www.cnn.com/2015/10/06/politics/special-operations-iron-man-talos-suit/

[61] Carafono, J. and Gudgel, A .(2007, 6 23). *National Security and Biotechnology: Small Science with a BigPotential*. Retrieved 9 14 2018, from The Heritage Foundation: http://www.heritage.org/defense/report/national-security-and-biotechnology-small-science-bigpotential

[62] Zolfagharifard, Z. (2016, 5 18). *London to New York in 35 minutes: Successful hypersonic Mach 7 engine test brings high-speed air travel a massive step closer to reality*. Retrieved 8 13 2017, from DailyMail.com:

http://www.dailymail.co.uk/sciencetech/article-3596220/Super-jet-technology-nears-reality-Australia-test.html

[63] *Nasa X-43*. Retrieved 10 5 2017, from Wikipedia: https://en.wikipedia.org/wiki/NASA_X-43

[64] Peplow, M .(2013, 11 20). *Graphene: The quest for supercarbon*. Retrieved 8 20 2017, from Nature: http://www.nature.com/news/graphene-the-quest-for-supercarbon-1.14193

[65] Silva, C (2017, 3 22). *The weapon of the future is here: New electromagnetic railgun can destroy targets 125 miles away*. Retrieved 4 23 2017, from Newsweek: http://www.newsweek.com/weapon-future-here-new-electromagnetic-railgun-can-destroy-targets-125-572205

[66] Ellis, R. (2015, 4 13). *Railgun Project takes 2001 concept into hands of U.S. Navy sailors*. Retrieved 4 6 2019, from Navy Live: https://navylive.dodlive.mil/2015/04/13/railgun-project-takes-2001-concept-into-hands-of-u-s-navy-sailors/

[67] *AN/SEQ-3 Laser Weapon System*. Retrieved 5 1 2017, from Wikipedia: https://en.wikipedia.org/wiki/Laser_Weapon_System

[68] Lengyel, G. (2007, 8 1). *Department of Defense Energy Strategy*. Retrieved 3 1 2018, from Brookings: https://www.brookings.edu/research/department-of-defense-energy-strategy/

[69] Condliffe, C. (2017, 3 2). *The Department of Defense wants to Double Down on Renewables*. Retrieved 1 16 2018, from MIT Technology Review: https://www.technologyreview.com/s/603778/the-department-of-defense-wants-to-double-down-on-renewables/

[70] Davis, S. and Lawrence, P. (1978). *Problems of Matrix Organizations*. Retrieved 2 12 2018, from Harvard Business Review: https://hbr.org/1978/05/problems-of-matrix-organizations

[71] Luo, I. (2015, 8 1). *7 Military Weapons China Copied from the United States*. Retrieved 7 10 2018, from The Epoch Times:

http://www.theepochtimes.com/n3/1699756-7-military-weap-ons-china-copied-from-the-united-states/

[72] US Naval Institute Staff (2015, 10 26). *China's Military Build with Cloned Weapons*. Retrieved 7 15 2017, from USNI News: https://news.usni.org/2015/10/27/chinas-military-built-with-cloned-weapons

[73] Mozur, P. and Perlez, J. (2017, 4 7). *China Tech Investment Flying Under the Radar, Pentagon Warns*. Retrieved 4 12 2017, from The New York Times: https://mobile.ny-times.com/2017/04/07/business/china-defense-start-ups-penta-gon-technology.html

[74] Marr, B. (2015, 2 25). *A brief history of big data everyone should read*. Retrieved 6 27 2017, from World Economic Forum: https://www.weforum.org/agenda/2015/02/a-brief-history-of-big-data-everyone-should-read/

[75] Carter, C. (2008, 9 16). *7 Ways to Foster Creativity in Your Kids*. Retrieved 4 6 2018, from Greater Good Magazine: http://greatergood.berkeley.edu/raising_happi-ness/post/7_ways_to_foster_creativity_in_your_kids

[76] Snyder, C. (2017, 5 4). *5 everyday inventions you didn't know came from DARPA*. Retrieved 1 16 2018, from Business Insider: http://www.businessinsider.com/5-inventions-darpa-gps-irobot-roomba-internet-2017-5

[77] Hempel, J. (2015, 11 18). *DoD Head Ashton Cater Enlists Silicon Valley to Transform the Military*. Retrieved 11 2 2017, from Wired: https://www.wired.com/2015/11/secretary-of-defense-ashton-carter/

[78] JASON (advisory group). Retrieved 8 17 2019, from Wikipedia: https://en.wikipedia.org/wiki/JASON_(advisory_group)

[79] Yanity, K. (2017, 3 17). *R.I. undersea innovation center connects companies with defense work*. Retrieved 3 17 2017, from

Providence Journal: http://www.providencejournal.com/news/20170317/ri-undersea-innovation-center-connects-companies-with-defense-work

[80] Weisgerber, M. (2014, 3 26). *Tech Giants Spend Billions more than Defense Firms on R&D*. Retrieved 4 6 2019, from Pakistan Defence: https://defence.pk/pdf/threads/tech-giants-spend-billions-more-than-defense-firms-on-r-d.316265/

[81] Dowdy, J. and Oakes, E.. *Defense outlook 2017: A global survey of defense-industry executives*. Retrieved 4 23 2018, from McKinsey and Company: http://www.mckinsey.com/industries/aerospace-and-defense/our-insights/defense-outlook-2017-a-global-survey-of-defense-industry-executives

[82] Balakrishnan, A. (2017, 4 7). *Apple CEO Cook says 'US will lose its leadership in technology' unless more women are hired*. Retrieved 10 14 2017, from CNBC: http://www.cnbc.com/2017/04/07/tim-cook-on-diversity-in-tech-at-auburn.html

[83] *Happy Hour*. Retrieved 4 20 2018, from Wikipedia: https://en.wikipedia.org/wiki/Happy_hour

[84] Finney, J. (1986, 7 9). Rickover, Father of Nuclear Navy, Dies at 86. Retrieved 1 20 2018, from The New York Times: https://www.nytimes.com/1986/07/09/obituaries/rickover-father-of-nuclear-navy-dies-at-86.html

[85] Beall, G. (2017, 11 6). 8 Key Differences between Gen Z and Millennials. Retrieved 12 5 2017, from HuffPost: http://www.huffingtonpost.com/george-beall/8-key-differences-between_b_12814200.html

[86] *National Security Decision Making*. Retrieved 8 26 2017, from U.S. Naval War College: https://www.usnwc.edu/Departments---Colleges/National-Security-Affairs/National-Security-Decision-Making-Course.aspx

[87] Cain, S. (2017, 3 24). *Not Leadership Material? Good. The World Needs Followers*. Retrieved 8 17 2017, from The New York Times: ttps://www.nytimes.com/2017/03/24/opinion/sunday/not-leadership-material-good-the-world-needs-followers.html?_r=0

[88] Williams, A. (2015, 9 18). *Move Over, Millennials, Here Comes Generation Z*. Retrieved 2018 1 13, from The New York Times: https://www.nytimes.com/2015/09/20/fashion/move-over-millennials-here-comes-generation-z.html?_r=0

[89] True Defective (@dubsteppenwolf) (2017, 5 17). *Good morning millennials. Let's take a look at all the things you murderous pricks have killed in the last two years, somehow*. Retrieved 4 6 2019, from Twitter: https://twitter.com/dubsteppenwolf/status/864902310006280193

[90] *Federal Spending: Where Does the Money Go*. Retrieved 2017 11 6, from National Priorities Project: https://www.nationalpriorities.org/budget-basics/federal-budget-101/spending/

[91] O'Hanlon, M. (2015, 8 19). *Dollars at work: What defense spending means for the U.S. economy*. Retrieved 12 12 2017, from Brookings: https://www.brookings.edu/blog/order-from-chaos/2015/08/19/dollars-at-work-what-defense-spending-means-for-the-u-s-economy/

[92] Alexander, R. (2012, 3 20). *Which is the world's biggest employer*. Retrieved 2018 2 16, from BBC: http://www.bbc.com/news/magazine-17429786

[93] Rizzo, J. (2011, 9 22). *Defense cuts: The jobs numbers game*. Retrieved 2018 3 13, from CNN: http://security.blogs.cnn.com/2011/09/22/defense-cuts-the-jobs-numbers-game/

[94] *List of states with nuclear weapons*. Retrieved 2017 11 12, from Wikipedia: https://en.wikipe-dia.org/wiki/List_of_states_with_nuclear_weapons

[95]Marmer, M., Herrmann, B., Dogrultan, E., Berman, R. (2012, 3). *Startup Genome Report Extra on Premature Scaling*. Retrieved 4 5 2019, from Amazon S3: https://s3.amazo-naws.com/startupcompass-public/StartupGe-nomeReport2_Why_Startups_Fail_v2.pdf

[96] Asay, M. (2017, 4 27). *Developers: Want to make serious money? Don't work for a startup*. Retrieved 3 11 2018, from TechRepublic: http://www.techrepublic.com/article/developers-want-to-make-serious-money-dont-work-for-a-startup/

[97] Asay, M (2017, 4 27). *Developers: Want to make serious money? Don't work for a startup*. Retrieved 3 11 2018, from TechRepublic: http://www.techrepublic.com/article/developers-want-to-make-serious-money-dont-work-for-a-startup/

About the Author

Jeremy A. Shattuck has spent the past the 20 years building systems for the US Navy. He has a B.S. in Electrical Engineering, M.B.A, M.A. in International Relations, Naval War College Diploma, and M.S. Systems Engineering from the Naval Postgraduate School, and is a recipient of the National Defense Industrial Association Ferguson Award for Systems Engineering Excellence.